Element	Symbol	Atomic Weight
Manganese	Mn	54.9380
Mercury	Hg	200.59
Molybdenum	Mo	95.94
Neon	Ne	20.183
Nickel	Ni	58.71
Nitrogen	N	14.0067
Osmium	Os	190.2
Oxygen	O	15.9994
Palladium	Pd	106.4
Phosphorus	P	30.9738
Platinum	Pt	195.09
Potassium	K	39.102
Rhenium	Re	186.2
Rhodium	Rh	102.905
Rubidium	Rb	85.47
Ruthenium	Ru	101.07
Selenium	Se	78.96
Silicon	Si	28.086
Silver	Ag	107.870
Sodium	Na	22.9898
Strontium	Sr	87.62
Sulfur	S	32.064
Tantalum	Ta	180.948
Tellurium	Te	127.60
Thallium	Tl	204.37
Thorium	Th	232.038
Tin	Sn	118.69
Titanium	Ti	47.90
Tungsten	W	183.85
Uranium	U	238.03
Vanadium	V	50.942
Xenon	Xe	131.30
Zinc	Zn	65.37
Zirconium	Zr	91.22

CHEMICAL
EQUILIBRIUM

HARPER'S CHEMISTRY SERIES
STUART A. RICE, EDITOR

CHEMICAL EQUILIBRIUM

Allen J. Bard

DEPARTMENT OF CHEMISTRY

THE UNIVERSITY OF TEXAS

HARPER & ROW, PUBLISHERS

NEW YORK, EVANSTON, AND LONDON

To the memory of
Bennie A. Ferrone

Preface

The theory of chemical equilibrium provides, for the beginning student, a good example of the development, application, and modification of a scientific theory. The conception of a theoretical model, the use of approximations and graphical methods for solving problems, and the modification of the original model to produce more accurate results, are familiar scientific operations. Because equilibrium theory is applicable to simple problems involving ionic solutions, it has become a traditional part of the introductory chemistry course. Unfortunately, teachers of more advanced courses often find that this initial exposure teaches students to solve problems by "cookbook" methods, so that problems only slightly different or more difficult require a relearning of the theory or memorization of still more formulas. This process sometimes continues through analytical, physical, and alas, even graduate courses.

This book is based on a belief that students of chemistry can appreciate a rigorous treatment of chemical equilibria right from the start, especially when approximations are explicitly introduced as definite steps in the solution of problems. The technique of solving problems by using material balance and electroneutrality equations employed in this book has been used before, but rarely in elementary treatments. Once a student masters this approach, he can at least embark upon the solution of any equilibrium problem without recourse to memorized formulas, and will hopefully retain this method from course to course.

The Brønsted approach to acid-base chemistry, which allows unification and simplification of problems and easy extension to nonaqueous systems, is used throughout. Although activity coefficients and the Debye-Hückel theory are usually not treated in introductory courses, the concepts are not beyond the grasp of most students, and hopefully, room for these topics will be found in already crowded courses. Chapters are also included on graphical methods, which provide an elegant approach to solving complicated problems, and on separations, a topic which is included in most analytical chemistry courses. Supplementary reading lists include references mainly to monographs and to papers in the *Journal of Chemical Education,* with the hope that at least a few students will become familiar with the chemical literature in this easily digestible form. These works will usually refer students to the original and more erudite papers not listed here.

The student is urged to work through the numerous example problems in the text, and also the problems at the end of the chapters (the answers to the (a) parts of most unsolved problems are given at the end of the book). Only by solving many problems can mastery of the subject be attained.

Many of my colleagues, especially Professor G. H. Ayres, and students have helped shape the viewpoint in this book through stimulating and critical discussions. I wish to thank Professor Fred Anson for reading the manuscript and making many helpful suggestions. My wife, Fran, typed the final draft of the manuscript and helped in proofreading; her constant encouragement and assistance is gratefully acknowledged.

Austin, Texas

ALLEN J. BARD

Contents

CHEMICAL
EQUILIBRIUM

CHAPTER 1 INTRODUCTION

1–1. THE USE OF EQUILIBRIUM THEORY

A chemist plans to study a reaction, such as the dissociation of a weak acid,

$$HF \rightleftarrows H^+ + F^- \tag{1.1}$$

the dissolution of a precipitate,

$$BaSO_4(\text{solid}) \rightleftarrows Ba^{++} + SO_4^{--} \tag{1.2}$$

an oxidation-reduction reaction,

$$2Fe^{+++} + 2I^- \rightleftarrows 2Fe^{++} + I_2 \tag{1.3}$$

or, in general, a reaction which can be represented as

$$aA + bB + \cdots \rightleftarrows cC + dD + \cdots \tag{1.4}$$

He wants to know whether the reaction will occur, to what extent it will proceed with different concentrations of the reactants, how the extent of the reaction will be affected by changes in temperature and pressure, etc. Obviously, he can obtain this information by going into his laboratory and carrying out a number of experiments and measurements under all possible conditions of interest. For example, if he wants to study the dissociation of HF in water at 25° C, with different mixtures of HF and F^-, he can make up solutions of 1 M, 0.5 M, 0.1 M, 0.01 M, and 0.001 M HF and measure the hydrogen ion concentration in each. Then he might prepare similar solutions with different concentrations of F^-, and again make the measurements. Such a direct approach requires the chemist to make a large number of measurements and in the process use up chemicals, glassware, and most importantly, time.

It is useful for a chemist to have a theory which allows him to perform relatively few experiments on his system, and then, to use the results of these few experiments coupled with his theory to predict the properties of the same system under different conditions. The Theory of Chemical Equilibrium is such a theory. Although experimentation is never completely eliminated by this theory, the approach is systematized so that only a few experiments are necessary. Frequently these experiments will already have been performed and the needed data recorded in some journal or book. Ideally, one would like to be able to calculate this data without recourse to any direct experiments, using only atomic properties or even the more fundamental properties of electrons and protons. This is the aim of theoretical chemistry. The nature of the mathematical and physical problems in this field are of such magnitude that the ability to predict the course of a reaction simply from fundamental properties, with a paper, pencil, and calculating machine, lies in the distant future for most reactions. The Theory of Chemical Equilibrium, based upon experimental measurements, will remain a useful tool for the chemist.

The Theory of Chemical Equilibrium is most fundamentally approached by consideration of the energy of a system (a *thermodynamic* approach); however, a clearer picture of a system proceeding to and reaching equilibrium can be obtained by considering the rates of the reactions (a *kinetic* approach). Both of these approaches will now be used to formulate the theory.

1–2. THE KINETIC APPROACH

THE FORWARD REACTION

Let us consider the formation of hydrogen iodide, HI, by the reaction of hydrogen, H_2, and iodine, I_2, in the gas phase

$$H_2 + I_2 \rightleftarrows 2HI \tag{1.5}$$

If we add, for example, 1 millimole of H_2 and 1 millimole of I_2 to a 1 liter flask, they will react to produce HI. The rate of the production of HI is proportional to the concentrations of hydrogen and iodine. Expressing this mathematically we have

$$v_f = k_f[H_2][I_2] \tag{1.6}$$

where v_f is the velocity of the forward reaction, k_f is simply a proportionality constant, and the brackets denote concentrations.

We can intuitively rationalize the above observations by the following argument: The reaction of H_2 and I_2 (or indeed most reactions) takes place because the molecules collide and atomic rearrangements occur. The greater the number of collisions per second between H_2 and I_2 molecules, the higher the rate of the production of HI; that is, the higher the rate of the reaction will be. Envision an experiment in which we place only a single molecule of I_2 and a single molecule of H_2 in a 1 liter flask (i.e., exceedingly small concentrations of I_2 and H_2). It will be a long time before these two molecules collide and thus the rate of the reaction will be very low. The opposite case would be placing a very large number of I_2 and H_2 molecules in a very small flask (i.e., high concentrations of I_2 and H_2). Under these conditions collisions will be more frequent and the rate of the reaction will be high. (1.6) simply states this fact. An investigation of this reaction has shown that at $527°$ C k_f is 2.3 liter/mole-second. Using this value and some calculus (see Appendix A), we can determine v_f and the concentrations of H_2, I_2, and HI. Results of these calculations are shown in Figure 1.1, (a) and (b).

THE BACKWARD REACTION

As the reaction proceeds, H_2 and I_2 are used up so that their concentrations continually decrease and v_f decreases. At the same time HI is produced and its concentration increases. However, as soon as HI molecules are produced they are free to collide, and in doing so, may sometimes break down again to form H_2 and I_2. The velocity of this backward reaction, v_b, was found experimentally to be proportional to the square of the concentration of HI, or

$$v_b = k_b[HI][HI] = k_b[HI]^2 \tag{1.7}$$

The concentration of HI is squared in this case because, for a reaction to occur, two molecules of HI collide. The value of k_b under these conditions was found to be 0.14 liter/mole-second, and v_b can be determined at any time (Fig. 1.1(b)). The velocity of the backward reaction increases

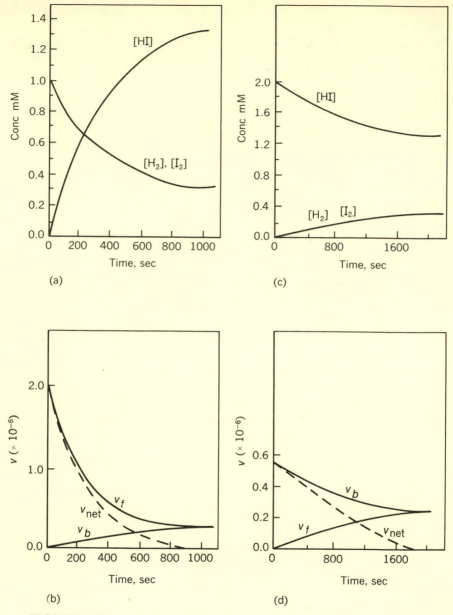

FIGURE 1.1. Reaction system: $H_2 + I_2 \rightleftarrows 2HI$. (a) Change in the concentrations of H_2, I_2, and HI during the reaction, when the system initially contains a $[H_2] = [I_2] = 0.001$ mole/liter, and $[HI] = 0$. (b) Changes in the velocities of the forward, v_f, and backward, v_b, reactions, with initial conditions of (a). (c) Change in the concentrations of H_2, I_2, and HI during the reaction, when the system initially contains a $[HI] = 0.002$ mole/liter, and a $[H_2] = [I_2] = 0$. (d) Changes in the velocities of the forward and backward reactions, with initial conditions of (c). See Appendix A for calculations.

continually during the course of the reaction because the concentration of HI continually increases. The overall reaction rate which we observe, v_{net}, is the difference between the velocities of the forward and backward reactions

$$v_{net} = v_f - v_b \qquad (1.8)$$

DYNAMIC EQUILIBRIUM

Since v_f decreases and v_b increases during the reaction, a point is eventually reached at which v_f and v_b are equal, or the overall reaction rate is zero. At this stage the system is said to be at *equilibrium*. No further overall changes in the concentrations of H_2, I_2, and HI will take place. In this system at this temperature equilibrium is reached after about 900 seconds. Although no apparent change is occurring in the reaction flask, both reactions are actually still proceeding; but, in opposite directions at the same rate. This point is often emphasized by calling this a *dynamic equilibrium*. Since v_f equals v_b at this point, then

$$k_f[H_2]_{eq}[I_2]_{eq} = k_b[HI]_{eq}^2 \qquad (1.9)$$

$$\frac{[HI]_{eq}^2}{[H_2]_{eq}[I_2]_{eq}} = \frac{k_f}{k_b} = K \qquad (1.10)$$

K is simply a number, a constant for this reaction at a given temperature, and is called the *equilibrium constant* for the reaction. Since no assumptions were made concerning any particular concentrations of H_2, I_2, and HI in writing (1.9) and (1.10), they should apply for any concentration of reactants of this system at a given temperature. From the values of k_f and k_b we can calculate $K = 16.4$ at 527° C. If we start with 1 mole/liter concentrations of H_2 and I_2, the rate at which equilibrium is approached will be different. Once equilibrium is reached, however, the concentrations of HI, H_2 and I_2, satisfy the equilibrium constant expression (1.10), with K equal to 16.4. This illustrates the usefulness of the equilibrium constant concept. Once K is determined by any one of a number of experimental methods for one set of concentrations in a given system, then this same K may be used to predict the final concentrations in this system in any other case at the same temperature.

(1.10) also makes no assumptions about the direction from which equilibrium is approached. If, in another experiment, we add 2 millimoles of HI to a 1 liter flask, it will decompose to form H_2 and I_2. The system is again described by the same set of equations, and the concentrations of the species and the reaction velocities may again be calculated

(Appendix A). The results of these calculations are found in Fig. 1.1, (c) and (d). In this case the system attains equilibrium in about 1800 seconds, but at this point the concentrations of H_2, I_2, and HI are exactly the same as in the preceding case. When these concentrations are put into the equilibrium constant expression they will yield the same numerical value of K.

COMPLEX REACTION MECHANISMS

These same ideas can be applied to a general reaction scheme, (1.4), and an equilibrium constant expression such as

$$\frac{[C]^c[D]^d \cdots}{[A]^a[B]^b \cdots} = K \tag{1.11}$$

can be derived. For many reactions the actual reaction mechanism may be more complicated than that of the H_2—I_2 reaction. For example it is not necessarily true that in a reaction

$$A + B \rightleftarrows 2C \tag{1.12}$$

the rate of the forward reaction is given by

$$v_f = k_f[A][B] \tag{1.13}$$

The reaction may actually occur in two steps

$$(1) \quad A + X \rightleftarrows AX \tag{1.14}$$

$$(2) \quad AX + B \rightleftarrows 2C + X \tag{1.15}$$

Here X is a third substance in the reaction mixture and is neither produced nor consumed in the overall reaction. In this case the rates of the individual reactions are given by

$$(1) \quad v_{1f} = k_{1f}[A][X] \qquad v_{1b} = k_{1b}[AX] \tag{1.16}$$

$$(2) \quad v_{2f} = k_{2f}[AX][B] \qquad v_{2b} = k_{2b}[C]^2[X] \tag{1.17}$$

However, once the system attains equilibrium, the rates of all forward and backward reactions are equal, and the expected equilibrium constant expression results

$$\frac{[C]^2_{eq}}{[A]_{eq}[B]_{eq}} = K = \frac{k_{1f}k_{2f}}{k_{1b}k_{2b}} \tag{1.18}$$

Although the actual mechanism of this reaction was more complex than the equation for the overall reaction (1.12), suggested, the resulting equilibrium constant expression is the same as that predicted by a simpler mechanism. This is a useful property of systems at equilibrium. No matter how complicated the reaction mechanism may be, the properties of the system at equilibrium are independent of the path taken from reactants to products, and the equilibrium constant expression can be written on the basis of the overall reaction itself.

In practice, equilibria are rarely studied by the method just described. The rates of many reactions are difficult to measure because they are either extremely high or extremely low, or because the reaction path is very complicated. The numerical values of the equilibrium constant can be determined more easily on a system "at rest"; for example, by measuring the concentrations of the reactants at equilibrium. This technique is usually used since it is easier to examine a system of constant composition rather than one which is changing with time.

1–3.   THE THERMODYNAMIC APPROACH

MECHANICAL SYSTEMS

Another way of defining the position of equilibrium is one based on energy considerations. A position of equilibrium is attained for a system when its energy is at a minimum. As a mechanical analogy of a chemical reaction, consider a ball resting on the side of a hill (Fig. 1.2). The ball,

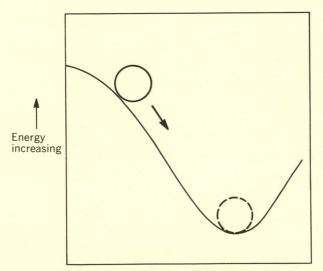

FIGURE 1.2.   A mechanical analogy to a chemical system. The ball on the hill will move until it reaches a position of minimum energy.

our system of interest, has a certain potential energy in this situation by virtue of its position on the hill. To lose this energy and attain a position of lower energy, the ball will roll down the hill and come to rest in the valley, where it attains the position of minimal energy for this system.

CHEMICAL SYSTEMS AND FREE ENERGY

A chemical system is very similar to this mechanical one of the ball. A given mixture of substances has a certain amount of energy associated with it (such as the kinetic energy of motion of the molecules, the bond energy holding atoms together, etc.). If this system can change by chemical reaction and form a new system with a lower energy, it will do so. The substances will react until a position of minimum energy is attained. Quantitatively, we can say that the total energy of a system containing n_A moles of substance A, n_B moles of substance B, etc., is

$$G_{total} = n_A G_A + n_B G_B + n_C G_C + \cdots \qquad (1.19)$$

G_{total} is the total free energy of the system, G_A is the free energy per mole of substance A, G_B is the free energy per mole of B, etc.

FREE ENERGY AND EQUILIBRIUM

Consider the reaction

$$A_2 + B_2 \rightleftarrows 2AB \qquad (1.20)$$

Assume 1.0 mole of A_2 and 1.0 mole of B_2 are added to a 1 liter flask. The free energy of the system will be

$$G = (1.0)G_{A_2} + (1.0)\ G_{B_2} \qquad (1.21)$$

After 0.1 mole of A_2 and 0.1 mole of B_2 have reacted to form 0.2 moles of AB, the free energy of the system will be

$$G' = (0.9)G'_{A_2} + (0.9)G'_{B_2} + (0.2)G'_{AB} \qquad (1.22)$$

We can determine the free energy of the system at any stage of this reaction by calculating the moles of each species present and the free energy of each species present in the mixture; details of these calculations are given in Appendix B. Typical results of calculations such as these for the hypothetical reaction (1.20) are shown in Figure 1.3. The energy of the system gets lower and lower as more AB is produced and A_2 and B_2 are consumed up to a certain point (indicated by the arrow). After this point, further production of AB and loss of A_2 and B_2 leads to an increase in the total free energy of the system. The position of equilibrium is the position of minimum energy for the system. The

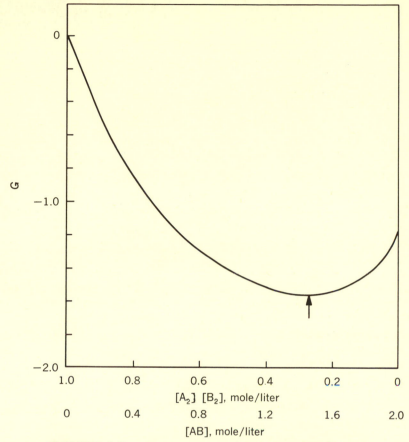

FIGURE 1.3. Change in the total free energy, G, of a system during different stages of the reaction $A_2 + B_2 \rightleftarrows 2AB$. At equilibrium the system will have minimum energy and a composition indicated by the arrow.

equilibrium constant for the reaction can be determined by putting the concentrations of A_2, B_2, and AB present at this point into the equilibrium constant expression. For this reaction at 25° C, K is 29.3. Obviously K is intimately related to the standard free energies of the components of a reaction (Appendix B).

METASTABLE EQUILIBRIA

The term *metastable* denotes a system at equilibrium that is not the most stable equilibrium possible for the system under those conditions. For example, consider the energy diagram for a system (Figs. .

1.2 and 1.3) and imagine a ridge or small valley on the side of the hill. The system, in proceeding to equilibrium, may stop in this small valley and be in a state of metastable equilibrium. This state is termed metastable because a more stable equilibrium, at the bottom of the hill, is available to it.

1–4. EQUILIBRIUM VS. KINETICS

EQUILIBRIUM CONSTANT EXPRESSIONS

Since the preceding discussions of equilibrium, from both the kinetic and thermodynamic point of view, involved consideration of a system approaching equilibrium, it may be well to stress here that the equilibrium constant expression is *only concerned with the final state of the system.* Given the equation for the overall chemical reaction, we can immediately write the corresponding equilibrium constant expression: Take the concentrations of the products of the reaction (on the right side of the equation for the reaction), raise them to a power of their respective coefficients in the chemical equation, and divide by the concentration of the reactants of the reaction (on the left side of the equation), each of these raised to a power of their respective coefficients. Then equate this expression to a constant. Table 1.1 contains the equilibrium constant expressions for several different types of reaction. Since we will be concerned only with chemical systems at equilibrium, that is systems in which the concentrations of the various reactants

TABLE 1.1. Equilibrium Constant Expressions for Various Chemical Reactions

Reaction type	Typical reaction	Equilibrium constant expression	Dimensions of K
General reaction	$aA + bB + \cdots \rightleftarrows cC + dD + \cdots$	$K = \dfrac{[C]^c[D]^d \cdots}{[A]^a[B]^b \cdots}$	
Ionization of a weak acid	$HF \rightleftarrows H^+ + F^-$	$K = \dfrac{[H^+][F^-]}{[HF]}$	mole/liter
Formation of a complex ion	$Ag^+ + 2NH_3 \rightleftarrows Ag(NH_3)_2{}^+$	$K = \dfrac{[Ag(NH_3)_2{}^+]}{[Ag^+][NH_3]^2}$	$(mole/liter)^{-2}$
Oxidation-reduction	$2Fe^{+++} + 2I^- \rightleftarrows 2Fe^{++} + I_2$	$K = \dfrac{[Fe^{++}]^2[I_2]}{[Fe^{+++}]^2[I^-]^2}$	$(mole/liter)^{-1}$

are not changing with time, we will henceforth use brackets without any subscript to represent concentrations of substances at equilibrium.

RATES OF REACTIONS

We will not be concerned with how long it takes a system to reach equilibrium. This is the concern of a considerably more complex subject called *chemical kinetics*. In all of our problems we will assume that we have waited for a time long enough to allow the system to come to "rest" (i.e., dynamic equilibrium). The time that may actually be required for a system to attain equilibrium may be either very short or very long. For the reaction

$$H^+ + OH^- \rightleftarrows H_2O \tag{1.23}$$

the equilibrium constant expression, with the correct numerical value of K at 25° C is

$$\frac{[H_2O]}{[H^+][OH^-]} = 5.6 \times 10^{15} \tag{1.24}$$

The rate of this reaction has been measured and is extremely high; the system essentially reaches equilibrium in less than a microsecond (one-millionth of a second). For a reaction which at first glance looks similar

$$2H_2 + O_2 \rightleftarrows 2H_2O \tag{1.25}$$

the equilibrium constant expression is

$$\frac{[H_2O]^2}{[H_2]^2[O_2]} = 10^{83} \; (25° \text{ C}) \tag{1.26}$$

The rate at which this reaction attains equilibrium under ordinary conditions is exceedingly low. If H_2 and O_2 gas are mixed in a flask and the contents of the flask are examined ten years later, a measureable amount of water will not be found. At equilibrium, (1.26) implies that essentially complete conversion of the H_2 and O_2 to water occurs. However, the system takes a long time to reach equilibrium. If finely divided platinum, which acts as a *catalyst* for this reaction, is added to the flask, the reaction will reach equilibrium very rapidly. A catalyst is a substance which alters the *rate* at which a system proceeds to equilibrium, but does not affect the final *position* of equilibrium. The reaction of H_2 and O_2 will also occur rapidly (explosively!) if a lighted match is put into the flask.

In summary, knowledge of the equilibrium constant gives no

information about the rate at which a system attains equilibrium. A system may have a very large equilibrium constant and attain equilibrium at a very low rate. Conversely, a reaction may have a small equilibrium constant and attain equilibrium rapidly.

1–5. DISPLACEMENT OF EQUILIBRIUM

In this section we are concerned with how the application of a change in conditions (a *stress*) to a system at equilibrium will change the concentrations of the reacting species. For example, suppose we have a system of H_2, I_2, and HI at equilibrium. At 527°C the concentrations must satisfy the equation

$$\frac{[HI]^2}{[H_2][I_2]} = 16.4 \qquad (1.27)$$

Suppose additional I_2 (or H_2 or HI) is added, or the pressure is increased, or the temperature is changed. How will the system react and what will be the new concentrations at equilibrium?

LE CHATELIER'S PRINCIPLE

One of the first investigators who attempted to answer these questions was Henry Le Chatelier. He deduced the following principle: When a change is made in a variable that determines the state of a system at equilibrium, the system will readjust in a manner as to counteract the change in the variable under consideration. This principle is sometimes stated: When a *stress* is applied to a system at equilibrium, the *position of equilibrium* will shift in such a direction as to counteract or undo the stress. For example, the system H_2, I_2, and HI, at equilibrium is represented by the chemical equation

$$H_2 + I_2 \rightleftarrows 2HI + \text{heat} \qquad (1.28)$$

(the heat term indicates that heat is evolved when H_2 and I_2 react). If additional I_2 is added to the system at equilibrium at 527° C, (1.27) must still be satisfied at the new position of equilibrium. At the instant the I_2 is added, the ratio of $[HI]^2$ to $[H_2]$ $[I_2]$ will be smaller than 16.4 because of this additional I_2. The system must adjust to make the denominator smaller and the numerator larger until (1.27) is satisfied. In other words, H_2 and I_2 will react and produce more HI until a new position of equilibrium is attained. It is often said that the addition of I_2 "shifts the equilibrium to the right." This terminology is satisfactory if one bears in mind that this simply means that H_2 and I_2 will react to form HI (or the "right" direction of (1.28)). An equilibrium system has no direction. Coming across an equilibrium system which contains various concentrations of both reactants and products, there is no way of predicting where the original conditions of this system arose from, and a literal interpretation of terms such as "shifts the direction of equilibrium ..." is meaningless.

TEMPERATURE CHANGES

Suppose the equilibrium mixture at 527° C is cooled to 283° C. Le Chatelier's principle predicts that the system will attempt to conteract this chilling by producing heat. Again H_2 and I_2 will react to form HI, since this is the direction of (1.28) which involves the production of heat. The change in the position of equilibrium in this case is reflected in changes in K, the equilibrium constant. At 283° C, K for (1.28) is 63, so that the production of HI from reactants is "more complete." Altering the position of equilibrium by changing the temperature is useful in causing reactions to proceed in the direction we desire. If the temperature is raised, the H_2, I_2, and HI, system will "shift to the left," or move in such a direction that heat is absorbed. More HI will break down to form H_2 and I_2. The effect of temperature changes on a system at equilibrium contrasts with the effect of such changes on reaction *rates*. While an equilibrium constant can be made either smaller or larger depending upon whether heat is evolved or absorbed during reaction, rate constants are almost always increased by increases in temperature. This increase in the rate of reaction is caused by the speeding up of the motions of the molecules which increases the number and energy of collisions of the molecules per unit time. Increasing the temperature will cause the rate constants of both the forward and backward reactions of (1.28) (k_f and k_b) to increase. Since (1.28) becomes less complete at higher temperatures however, and $K = k_f/k_b$, k_b increases more than k_f in this case.

PRESSURE CHANGES

The effect of pressure changes on a system at equilibrium depends upon the particular reaction and whether the reaction is occurring in the gas phase or in solution. In gaseous reactions, pressure changes will affect the position of equilibrium if the total number of moles of gas in the system changes during the reaction of interest. For example, in the reaction

$$N_2O_4 \rightleftarrows 2NO_2 \qquad (1.29)$$

one mole of N_2O_4 produces two moles of NO_2. Since one mole of any gas occupies about the same volume (for an ideal gas this is 22.4 liters per mole at $0°$ C and 760 mm Hg pressure), this means that (1.29) occurs with an increase in volume at constant pressure. By Le Chatelier's principle, an increase in pressure will cause the system to react in a manner as to undo this stress, namely by decreasing the volume. The equilibrium position will change to one with a higher concentration of N_2O_4 and a lower concentration of NO_2 (or the reaction will "shift to the left"). For gaseous reactions which occur without any change in the total number of moles of gas, for example, (1.28), the position of equilibrium is unaffected by pressure changes.

Generally pressure affects equilibria in solution in the same way, but to a much smaller extent, since liquids are much less compressible than gases. For reactions which do occur with a change in volume at constant pressure, an increase in pressure will cause the equilibrium condition to change to one where reaction occurs in the direction of a decrease in volume. Reactions that involve the formation of ions from molecules (such as ionization), or the formation of highly charged ions from ions of lower charge, generally take place with a decrease in volume. For example, the following reactions occur with a net decrease in volume, and the position of equilibrium would "shift to the right" in all of the cases with increasing pressure:

$$H_2O \rightleftarrows H^+ + OH^-$$

$$FeCl^{++} \rightleftarrows Fe^{+++} + Cl^-$$

$$BeSO_4 \rightleftarrows Be^{++} + SO_4^{--}$$

The net change in the equilibrium constant is quite small however. For example, for the ionization of water, the equilibrium constant just about doubles when the pressure is increased from 1 to 1000 atmospheres at room temperature.

A recent interesting application of displacement of equilibria involves its use in the measurement of rates of very fast reactions. If a reaction occurs very rapidly, it is difficult to study by the most straightforward approach of mixing the reactants and observing the changes in concentrations with time, because the reaction will be essentially over before the reactants are completely mixed. What is done in these cases is to take a system already at equilibrium, and cause it to be displaced from its equilibrium position by changing, very suddenly, either the temperature (the *temperature jump* method) or the pressure (the *pressure jump* method), and watching the rate at which the system attains its new position of equilibrium.

1–6. THE NUMERICAL VALUES OF EQUILIBRIUM CONSTANTS

Equilibrium constants are often determined by allowing a system to attain equilibrium and then measuring the concentrations of the species in the mixture, although other, less direct, methods can also be used. Tables of numerical values of equilibrium constants for different types of reactions are given in Appendix C. These equilibrium constants will be used in solving the problems in the following chapters.

The equilibrium constant expression is a specific relationship involving concentrations. Concentration has the dimensions (or units) of amount of material per unit volume, such as grams/liter, moles/liter, etc. For the same reaction, the value of the equilibrium constant will depend upon the units in which concentration is expressed. By convention, since we all want to be able to use the same tables of equilibrium constants with a minimum of difficulty, we will use the units usually used in solution equilibria expressions, *moles/liter*, so that these equilibrium constants, in general, have dimensions in moles/liter (Table 1.1). Tabulated equilibrium constants for ionic equilibria are usually given without these dimensions, however. *In solving problems, it will always be necessary to convert all data to the units of moles/liter before attempting to use them in the equilibrium constant expressions.* Unless noted otherwise, all reactions in the discussions that follow are assumed to be occurring at 25° C (room temperature).

SUPPLEMENTARY READING

Butler, J. N., *Ionic Equilibrium*, Reading, Mass.: Addison-Wesley, 1964.

Carnell, P. H., and R. N. Reusch, *Molecular Equilibrium*, Philadelphia: Saunders, 1963.

Freiser, H., and Q. Fernando, *Ionic Equilibria in Analytical Chemistry*, New York: Wiley, 1963.

Frost, A. A., Effect of Concentration on Reaction Rate and Equilibrium, *J. Chem. Educ.*, **18**, 272 (1941).

Hammes, G., and L. E. Erickson, Kinetic Studies of Systems at Equilibrium, *J. Chem. Educ.*, **35**, 611 (1958). Application of the "pressure jump" and "temperature jump" methods to the study of rapid reactions.

de Heer, J., Principle of Le Chatelier and Braun, *J. Chem. Educ.*, **34**, 375 (1957).

Lee, T. S., in I. M. Kolthoff and P. J. Elving, *Treatise on Analytical Chemistry*, New York: Interscience, 1959, Vol. 1, Part I, p. 189.

Lindauer, M. W., The Evolution of the Concept of Chemical Equilibrium from 1775 to 1923, *J. Chem. Educ.*, **39**, 384 (1962). A very good discussion of the development of equilibrium theory.

Mysels, K. J., Laws of Reaction Rates and Equilibrium, *J. Chem. Educ.*, **33**, 178 (1956).

CHAPTER 2 CALCULATIONS WITH EQUILIBRIUM CONSTANT EXPRESSIONS

The problem facing the chemist in Chapter 1, finding the hydrogen ion concentration in different solutions containing HF and NaF, is typical of many equilibrium problems. Different substances are added to a reaction vessel and the concentrations of the various species at equilibrium are calculated. For example certain amounts of HF and NaF are added to a certain volume of water and the concentrations of H^+, Na^+, F^-, and HF* are calculated. Other related problems involve the determination of the amount of reagent needed to cause precipitation of a certain cation, prediction about the ability of a certain reagent to oxidize or reduce another substance, or determination of the extent of a reaction. All these problems involve the use of equilibrium constant expressions, certain other necessary equations, algebra, and, very often, some chemical intuition.

* Most species in aqueous solutions are associated with one or more molecules of water, i.e., are *hydrated*, so that species written as H^+, Na^+ ··· should more exactly be written $H(H_2O)_n{}^+$, $Na(H_2O)_m{}^+$, ··· (where n and m are *hydration numbers*.) For simplicity we will usually write the formulas of the bare ions and molecules with the understanding that these denote the hydrated species.

2–1. IONIZATION

Many equilibrium problems of interest in elementary work involve aqueous solutions. Water has as one property the ability to cause many substances to split apart into charged species, that is, to dissociate to form ions. Before attempting to solve equilibrium problems involving aqueous solutions, it is necessary to review what substances ionize in water and what substances do not. It is convenient to divide these substances into two categories: those that are essentially completely dissociated into ions in water (*strong electrolytes*) and those which are incompletely dissociated (*weak electrolytes*).

STRONG ELECTROLYTES

Salts

Almost all salts are completely dissociated in water. When 1.0 mole of NaCl is added to 1 liter of water, the resulting solution contains 1.0 mole per liter (or 1.0 M) Na^+ and 1.0 mole per liter (or 1.0 M) Cl^-, and essentially no species which could be considered undissociated NaCl. Other such salts include KBr, $Ca(NO_3)_2$, $Fe(ClO_4)_3$, and $Cd(BrO_3)_2$. The only apparent exceptions to the complete dissociation of salts are the complex species (e.g., $HgCl_2$, $CdBr_2$) or ion pair formation.*

Strong Acids

Only few acids are essentially completely dissociated and ionized in water. The most important ones are HCl, HNO_3, $HClO_4$, and H_2SO_4 (first step). Others include HBr, HI, and H_2SeO_4. Except for these, and it is recommended that the student memorize the first four, most other acids are weak. Some of these acids are completely ionized only in relatively dilute solutions (less than 1 M). For example, in concentrated solutions HNO_3 is predominantly in an un-ionized form. On the other hand $HClO_4$, HCl, and H_2SO_4 are more than 99 percent ionized in solutions with concentrations up to 10 M. Since we usually deal with relatively dilute solutions, for our problems we will consider all these acids as completely ionized.

---

* A negatively charged ion and a positively charged ion may sometimes be held together by electrostatic forces (forces arising from the attraction of opposite charges) in the form of an *ion pair*. The extent of ion pair formation in dilute aqueous solutions is usually slight and will be neglected in most of our discussions (see, however, p. 41 and Chapter 7).

Strong Bases

The important bases which are completely dissociated are NaOH, KOH, LiOH, and $Ba(OH)_2$.

Solving problems involving only strong electrolytes is generally easy, since no equilibria are involved. We can generally assume 100 percent dissociation and calculate the concentrations of the ions by stoichiometry.

Example 2.1. Calculate the concentrations of the ions in a solution prepared by diluting 0.10 mole of $Ca(NO_3)_2$, 0.30 mole of HCl, and 0.20 mole of $CaCl_2$ to 1 liter with water.

Since all three substances are strong electrolytes they are completely ionized. Therefore:

$$[H^+] \quad = 0.30 \text{ mole/l} = 0.30 \ M$$
$$[NO_3^-] = 2(0.10) = 0.20 \ M$$
$$[Ca^{++}] = (Ca^{++} \text{ from } Ca(NO_3)_2) + (Ca^{++} \text{ from } CaCl_2)$$
$$[Ca^{++}] = 0.10 + 0.20 = 0.30 \ M$$
$$[Cl^-] \quad = (Cl^- \text{ from HCl}) + (Cl^- \text{ from } CaCl_2)$$
$$[Cl^-] \quad = 0.30 + 2(0.20) = 0.70 \ M$$

Example 2.2. Show that the above solution, prepared by adding uncharged substances to water, is electrically neutral; that is, that the total "moles" of positive charge are equal to the total "moles" of negative charge.

Each univalent ion contributes one mole of positive or negative charge per mole, so that 0.30 mole of H^+ is equivalent to 0.30 mole of positive charge. For a divalent ion, Ca^{++} for example, each mole contributes two moles of charge, so that 0.30 mole of Ca^{++} per liter is equivalent to $2[Ca^{++}]$, or 0.60 mole, of positive charge per liter. On this basis we must show that

$$[H^+] + 2[Ca^{++}] = [Cl^-] + [NO_3^-] \tag{2.1}$$

Since $(0.30) + 2(0.30) = (0.70) + (0.20)$ the problem is solved, as required.

These examples illustrate two principles that we will use very often in solving problems. The first is the *material balance condition*, which requires that the total number of atoms of a given element introduced into a solution remains constant, no matter how these atoms may be redistributed among different species by reaction. If 1.0 mole of H_2S is added to a solution, we will find that it dissociates to some extent into HS^- and S^{--}. However the total amount of S in all of these species is still 1.0 mole. The second principle is the *electroneutrality* or *charge balance condition*, which requires that when uncharged substances are added to an uncharged solvent, the solution will remain electrically neutral. We will use both of these principles in solving equilibrium problems that involve weak electrolytes.

WEAK ELECTROLYTES

Weak Acids

Most acids, and almost all organic acids, are incompletely dissociated. These include H_2S, H_2CO_3, HF, HBO_2, CH_3COOH (acetic acid), C_6H_5COOH (benzoic acid), and many others.

Weak Bases

Weak bases, which react incompletely with water to form ions, include NH_3 (ammonia), N_2H_4 (hydrazine), and most organic bases, such as $C_6H_5NH_2$ (aniline), $C_2H_5NH_2$ (ethylamine), and others.

Complexes

Complexes or coordination compounds, species composed of a central metal ion and one or more associated groups, are incompletely dissociated. Typical complex ions, such as $Ag(NH_3)_2^+$, $Fe(CNS)^{++}$, $Hg(CN)_4^{--}$, $Fe(CN)_6^{---}$, and others are discussed in Chapter 5.

Water

Water itself is only very slightly ionized into H^+ and OH^-.

Others

Most soluble organic compounds are not ionized. These include sugar ($C_{12}H_{22}O_{11}$), ethyl alcohol (C_2H_5OH), and many others. The same is true for many dissolved gases, such as O_2, H_2, and N_2.

Problems involving weak electrolytes are sometimes difficult to solve, since they involve one or more dissociation equilibrium constant expressions. These problems will be treated later. The preceding list of strong and weak electrolytes must be understood and essentially committed to memory before problems in ionic equilibria can be undertaken.

2–2. SOLVING PROBLEMS

Generally equilibrium problems are solved by converting the given chemical data into mathematical equations. From algebra we remember that we must have one equation for each unknown which appears in the problem. Usually three types of equations are written in equilibrium problems: equilibrium constant expressions, charge balances, and material balances.

Example 2.3. What concentrations of H^+, F^-, and HF are present in a solution prepared by dissolving 1.00 mole of HF in water to give 1 liter of solution?

Step 1. Write chemical equations for all reactions that occur. Since HF is a weak acid, it only partially dissociates into ions (acid-base chemistry is discussed in Chapter 3)

$$HF \rightleftharpoons H^+ + F^- \tag{2.2}$$

Step 2. Write all pertinent equilibrium constant expressions. From (2.2) we can write (looking up the value of K_a in Appendix C)

$$\frac{[H^+][F^-]}{[HF]} = K_a = 6.7 \times 10^{-4} \tag{2.3}$$

This is the only equilibrium constant expression of interest in this problem (neglecting the ionization of water). In other problems several equilibria might be involved, and a number of equilibrium constant expressions would have to be written. (2.3) contains three unknowns, $[H^+]$, $[F^-]$, and $[HF]$, and two more equations involving these quantities must be supplied.

Step 3. Write the electroneutrality equation. Remembering that water is uncharged, and all substances added are uncharged (containing equal numbers of positive and negative charges), we can write the electroneutrality condition (see Example (2.2)). Since only H^+ and F^- are present in significant amounts, we can write

$$[H^+] = [F^-] \tag{2.4}$$

Step 4. Write the material balance equations. These equations must supply the additional equations, other than equilibrium constant expressions and the electroneutrality condition. The two equations written thus far, (2.3) and (2.4), would apply to any problem involving HF and water. We now make use of the fact that 1.00 mole of HF was added to water by noting that the total concentration of all fluoride-containing species in the solution must add up to the total amount of fluoride added, i.e., 1.00 mole, so that

$$[HF] + [F^-] = 1.00 \ M \tag{2.5}$$

Since hydrogen and oxygen are components of water, it is usually not fruitful to write material balance equations for these elements.

Step 5. Solve the equations algebraically. Now that we have three equations for our three unknowns the problem ceases to be a chemical problem and becomes an algebraic one. Solving (2.5) for [HF], and (2.4) for [F$^-$], and introducing these values into (2.3), we obtain

$$\frac{[H^+]^2}{1.00 - [H^+]} = 6.7 \times 10^{-4} \tag{2.6}$$

This may be rewritten as

$$[H^+]^2 + (6.7 \times 10^{-4})[H^+] - (6.7 \times 10^{-4}) = 0 \tag{2.7}$$

which is a quadratic (second degree) equation and may be solved by the quadratic formula.*

$$[H^+] = \frac{-(6.7 \times 10^{-4}) + \sqrt{(44.9 \times 10^{-8}) + 4(6.7 \times 10^{-4})}}{2}$$

$$[H^+] = 2.56 \times 10^{-2} \ M = 0.0256 \ M$$

From (2.4) and (2.5) we calculate that $[F^-] = 0.0256 \ M$ and $[HF] = 0.97 \ M$.

Frequently, we can eliminate much of the algebraic labor in problems of this type by using our chemical intuition and making some approximations. Since K_a for the dissociation of HF is fairly small, the extent of ionization of HF will be small, and we might guess that the concentration of H$^+$ in this solution will also be small compared

---

* This formula states that the equation

$$ax^2 + bx + c = 0$$

may be solved for x using the formula

$$x = \frac{-b \pm \sqrt{b^2 - 4ac}}{2a}$$

with the sign being chosen in front of the square root sign in equilibrium problems to yield a positive value of x.

to the concentration of undissociated HF, or will be small compared to 1.00 in (2.6), that is

$$[H^+] \ll 1.00$$

Whenever a number is added to or subtracted from another much larger number, we can as a first approximation, neglect the smaller number.* Hence we can say that 1.00 minus the hydrogen ion concentration is approximately equal to 1.0, or

$$1.00 - [H^+] \approx 1.0$$

Now (2.6) is much simpler, and the result is

$$[H^+] = \sqrt{6.7 \times 10^{-4}} = 2.59 \times 10^{-2} \ M$$

a value close to the one obtained without making any approximations.
Example 2.4. Calculate the concentrations of Ca^{++}, Ac^-, H^+, and HAc in a solution prepared by dissolving 1.0 mole of acetic acid, CH_3COOH (which we will abbreviate HAc, with CH_3COO^- being written Ac^-) and 1.0 mole of $CaAc_2$ in water to give 1.0 liter of solution.

We proceed as we did in Example 2.3, following steps (1) through (5):

(1) $$HAc \rightleftarrows H^+ + Ac^-$$

(2) $$\frac{[H^+][Ac^-]}{[HAc]} = K_a = 1.8 \times 10^{-5}$$

(3) $$[H^+] + 2[Ca^{++}] = [Ac^-]$$

(4) $$[Ca^{++}] = 1.0 \ M$$

$$[HAc] + [Ac^-] = 1.0 + 2(1.0) = 3.0 \ M$$

(5) Combination of the above equations leads to

$$\frac{[H^+](2.0 + [H^+])}{1.0 - [H^+]} = 1.8 \times 10^{-5}$$

The reader may verify that both the approximate solution (using the approximation $[H^+] \ll 1.0$) and the exact solution (which in this case is quite difficult) lead to the answer, $[H^+] = 9.0 \times 10^{-6} \ M$.

* That is, that $1.00 + 0.0001 \approx 1.00$. Note that this rule obviously does not apply to numbers *multiplied* or *divided* by a much smaller number.
Some mathematical symbols we will use are:
  $\ll$ is much smaller than
  $\gg$ is much larger than
  (note that the points always point at the smaller number).
  $\approx$ is approximately equal to

The general procedure outlined in this chapter will be useful for all types of equilibria, and in all kinds of problems. The difficulty of this approach is that for more complex problems the mathematical equations which must be solved are quite difficult, often involving third or higher degree equations. Sometimes we can solve problems more easily by using our chemical knowledge of the system. Both approaches will be taken in the succeeding chapters. A graphical method that is particularly useful in very complex problems, or for a rapid survey of a system, will be described in Chapter 8.

PROBLEMS

2.1. Calculate the concentration of each ion in the solutions.

   (a) 0.10 mole $BaCl_2$, 0.50 mole KNO_3, and 0.30 mole LiCl diluted to 500 ml

   (b) 33.6 grams $Ba(ClO_4)_2$ and 11.2 grams $Mg(ClO_4)_2$ diluted to 750 ml

   (c) 8.57 grams $Ba(OH)_2$, 0.10 mole $BaCl_2$, and 200 ml of 0.30 M NaOH diluted to 1 liter

   (d) 0.20 mole $Al_2(SO_4)_3$, 39.2 grams $FeSO_4 \cdot (NH_4)_2SO_4 \cdot 6H_2O$ and 40.0 grams $Fe_2(SO_4)_3$ diluted to 2 liters

2.2 Write the ionic balance (electroneutrality) condition for each of the solutions in problem 2.1 in terms of the concentrations of the ions present in these solutions. Demonstrate that the ionic balance equation is correct by substituting the numerical values of the concentrations found in problem 2.1.

2.3. List all of the species, ions and molecules, present in the following solutions. Write the electroneutrality condition and the material balance equations for each.

   (a) a solution prepared by diluting 0.10 mole HCN to 500 ml

   (b) a solution prepared by diluting 0.10 mole HCN, 0.20 mole HCl and 0.10 mole NaCN to 1 liter

   (c) a solution prepared by diluting 0.10 mole NaCN, 0.20 mole NaCl, and 0.30 mole HCN to 1 liter

   (d) a solution prepared by diluting 0.10 mole HCl, 0.20 mole NH_4Cl, and 0.20 mole HF to 500 ml

   (e) a solution prepared by diluting 0.10 mole H_2S and 0.10 mole HCl to 1 liter (Hint: H_2S produces H^+, HS^-, and S^{--} upon ionization.)

2.4. Combine the following equations and solve for x. Use the quadratic formula if necessary.

(a) $x + y + z = 10$
$x - z = 41$
$x - y = 30$

(b) $x + y = 5x$
$y = z + x$
$18 + 2x = z$

(c) $x + y = 0.600$
$xy = 0.0275$

(d) $\dfrac{xy}{z} = a$
$x + y = 2z$
$z + x = b$

(1) Solve in terms of a and b.
(2) Solve for the special case $a = 1$, $b = 2$.

2.5. Solve the following sets of equations for the concentrations of each species, (1) assuming $[H^+] \ll 0.01$ M, and (2) rigorously.

(a) $\dfrac{[H^+][X^-]}{[HX]} = 1.0 \times 10^{-4}$
$[H^+] = [X^-]$
$[HX] + [X^-] = 1.0$

(b) $\dfrac{[H^+][A^-]}{[HA]} = 1.0 \times 10^{-2}$
$[HA] + [A^-] = 3.0$
$[Na^+] = 1.0$

$[H^+] + [Na^+] = [A^-]$

(c) $\dfrac{[H^+][B^-]}{[HB]} = 1.0 \times 10^{-3}$
$[H^+] = [B^-]$
$[HB] + [B^-] = 0.010$

(d) $\dfrac{[H^+][X^-]}{[HX]} = 1.0 \times 1.0^{-5}$
$[H^+] + [K^+] = [X^-]$
$[K^+] = 0.10$

$[HX] + [X^-] = 0.50$

SUPPLEMENTARY READING

Butler, J. N., An Approach to Complex Equilibrium Problems, *J. Chem. Educ.*, **38**, 141 (1961).

Radimer, K. J., Solution of Problems Involving Equilibrium Constants, *J. Chem. Educ.*, **27**, 251 (1950).

CHAPTER 3 ACID-BASE EQUILIBRIA

3–1. ACIDS AND BASES

ACIDS

A useful definition of an acid is given by Brønsted and Lowry: an acid is a proton (i.e., hydrogen ion) donor.[*] A substance such as HF is an acid because it can donate a proton to a substance capable of accepting it. In aqueous solutions water is always available as a proton acceptor, so that the ionization of an acid, HA, can be written as

$$HA + H_2O \rightleftarrows H_3O^+ + A^- \qquad (3.1)$$

Since most of this chapter involves ionization of acids in aqueous solution, (3.1) will be written

$$HA \rightleftarrows H^+ + A^- \qquad (3.2)$$

Remember that H^+ really stands for a hydrated proton (a proton associated with one or more water molecules). The equilibrium constant

* J. N. Brønsted, *Rec. trav. chim.*, **42**, 718 (1923). T. M. Lowry, *J. Soc. Chem. Ind., London*, **42**, 43 (1923).

expression for (3.2), writing K_a as the equilibrium constant for the ionization of an acid, is

$$\frac{[H^+][A^-]}{[HA]} = K_a \qquad (3.3)$$

Examples of ionization of acids are:

$$HCl \rightarrow H^+ + Cl^- \qquad K_a \approx \infty \qquad (3.4)$$

$$HF \rightleftarrows H^+ + F^- \qquad K_a = 6.7 \times 10^{-4} \qquad (3.5)$$

$$NH_4^+ \rightleftarrows NH_3 + H^+ \qquad K_a = 5.5 \times 10^{-10} \qquad (3.6)$$

$$HCN \rightleftarrows H^+ + CN^- \qquad K_a = 7.2 \times 10^{-10} \qquad (3.7)$$

BASES

A base is defined as a proton acceptor. In aqueous solutions water is always available to donate a proton to a base, so the ionization of a base B, may be written

$$B + H_2O \rightleftarrows HB^+ + OH^- \qquad (3.8)$$

The equilibrium constant expression for the dissociation of a base, writing K_b as the equilibrium constant for the ionization of a base, is

$$\frac{[HB^+][OH^-]}{[B]} = K_b \qquad (3.9)$$

Examples of ionization of bases are:

$$CN^- + H_2O \rightleftarrows HCN + OH^- \qquad K_b = 1.4 \times 10^{-5} \qquad (3.10)$$

$$NH_3 + H_2O \rightleftarrows NH_4^+ + OH^- \qquad K_b = 1.8 \times 10^{-5} \qquad (3.11)$$

$$F^- + H_2O \rightleftarrows HF + OH^- \qquad K_b = 1.5 \times 10^{-11} \qquad (3.12)$$

The *strength* of an acid or base is related to the extent that the dissociation reactions proceed to the right, or to the magnitude of K_a or K_b; the larger the equilibrium constant for the dissociation reaction, the stronger the acid or base. From the values of K_a we can see that HCl is the strongest acid of those listed above (being essentially 100 percent ionized), followed by HF, HCN, and NH_4^+. From the K_b's given, NH_3 appears as the strongest base of those listed, followed by CN^- and F^-. Clearly, when an acid ionizes it produces a base. The acid, HA, and the base produced when it ionizes, A^-, are called a *conjugate acid-base pair*, so that the couples HF, F^- and HCN, CN^- and NH_4^+, NH_3, are conjugate acids and bases. Chloride ion is a negligibly weak base in water, since its conjugate acid, HCl, is essentially completely ionized and Cl⁻ has no tendency to combine with a proton.

AMPHOLYTES

A substance which can either lose or gain a proton, that is, behave like an acid or a base, is called an *ampholyte* (or an *amphiprotic* substance). Water is an ampholyte, since it behaves as a base in the presence of acids, (3.1), and as an acid in the presence of bases, (3.8).

WATER

Water itself can ionize according to the reaction

$$H_2O + H_2O \rightleftarrows H_3O^+ + OH^- \tag{3.13}$$

or in an abbreviated form

$$H_2O \rightleftarrows H^+ + OH^- \tag{3.14}$$

The equilibrium constant for this reaction at 25° C is given by the expression

$$\frac{[H^+][OH^-]}{[H_2O]} = K = 1.8 \times 10^{-16} \tag{3.15}$$

This equation can be simplified by noting that the concentration of water, $[H_2O]$, in *dilute* aqueous solutions is approximately constant and equal to

$$\frac{1000 \text{ g/liter}}{18 \text{ g/mole}} = 55.5 \text{ moles/liter}$$

(3.15) may then be written

$$[H^+][OH^-] = 55.5 \times 1.8 \times 10^{-16} = 1.0 \times 10^{-14} = K_w \tag{3.16}$$

In fact, the water concentration term is usually omitted from the equilibrium constant expression in any equilibrium occurring in dilute aqueous solution with water as a reactant or product, and $[H_2O]$ "lumped" with the equilibrium constant, K, as above. For example, this procedure was followed in writing both the K_a expression, (3.3), and the K_b expression, (3.9).

(3.16) will always hold in dilute aqueous solutions, and we can always write this in addition to any other equilibrium constant expressions when it is useful for us to do so. A solution containing only water will satisfy the ionic balance equation

$$[H^+] = [OH^-] \tag{3.17}$$

Combining (3.16) and (3.17), we find that a neutral solution is one in which

$$[H^+] = [OH^-] = \sqrt{K_w} = 1.0 \times 10^{-7} \; M \qquad (3.18)$$

K_w, the *ion product constant* for water, is a function of temperature, as are most other equilibrium constants. For example at $60°$ C, K_w is about 10^{-13}, and a neutral solution at this temperature would have a $[H^+]$ of $3.2 \times 10^{-7} \; M$. Solutions in which the hydrogen ion concentrations are larger than the hydroxyl ion concentrations are acidic solutions, while solutions with higher hydroxyl ion concentrations are basic solutions. Except for solutions where the $[H^+]$ has a value between $1.0 \times 10^{-6} \; M$ and $1.0 \times 10^{-8} \; M$, either H^+ or OH^- is large enough in excess so that it may be useful to neglect one or the other of these in doing calculations.

pH

It is sometimes convenient to specify the logarithm of a concentration or equilibrium constant. Generally this is given the notation pX, which is defined

$$pX = - \log X \qquad (3.19)$$

We have, therefore

$$pH = - \log [H^+]* \qquad (3.20)$$

$$pOH = - \log [OH^-] \qquad (3.21)$$

and taking logarithms of (3.16)

$$- \log [H^+] - \log [OH^-] = -14.00 \qquad (3.22)$$

or

$$pH + pOH = 14.00 = pK_w \qquad (3.23)$$

Therefore in a neutral solution at $25°$ C

$$pH \, p = OH = 7.00 \qquad (3.24)$$

Example 3.1. Calculate the pH, pOH, and $[OH^-]$ in a solution in which $[H^+] = 3.0 \times 10^{-4} \; M$.

$$pH = - \log (3.0 \times 10^{-4}) = - \log - 3.0 \log (10^{-4})$$

$$= -0.48 - (-4.00) = 3.52 \quad \textit{Answer}$$

$$pOH = 14.00 - 3.52 = 10.48 \quad \textit{Answer}$$

---

* pH is actually defined in terms of measurements on standard solutions, and fits more closely the equation pH $= - \log a_{H^+}$, where a_{H^+} is the *activity* of hydrogen ion (see Chapter 7).

$$[OH^-] = \frac{1.0 \times 10^{-14}}{3.0 \times 10^{-4}} = 3.3 \times 10^{-11} \ M \quad \textit{Answer}$$

Example 3.2. What is the $[H^+]$ in a solution of $pH = 10.70$?

$$\log [H^+] = -pH = -10.70 = -11.00 + 0.30*$$

$$[H^+] = \text{antilog} \ (-11.00) + \text{antilog} \ (0.30)$$

$$[H^+] = 10^{-11} \times 2.0 \ M \quad \textit{Answer}$$

	pH	[H+] M	[OH−] M	pOH	
	14.0			0.0	
	13.0	10^{-13}	10^{-1}	1.0	
	12.0			2.0	
Basic	11.0	10^{-11}	10^{-3}	3.0	$[OH^-] \gg [H^+]$
	10.0			4.0	
	9.0	10^{-9}	10^{-5}	5.0	
	8.0			6.0	
Neutral	7.0	10^{-7}	10^{-7}	7.0	$[H^+] \sim [OH^-]$
	6.0			8.0	
	5.0	10^{-5}	10^{-9}	9.0	
	4.0			10.0	
Acidic	3.0	10^{-3}	10^{-11}	11.0	$[H^+] \gg [OH^-]$
	2.0			12.0	
	1.0	10^{-1}	10^{-13}	13.0	
	0.0			14.0	
	−1.0	10	10^{-15}	15.0	

FIGURE 3.1. Relations between [H+], [OH−], pH, and pOH in aqueous solutions at 25° C.

* -10.70 is written in this way, rather than as $-10.00 - 0.70$, because log tables only list positive mantissas, and it is easier to find the antilog of 0.30 than it is to find the antilog of -0.70.

The pH notation implies that the higher the $[H^+]$ (the more acidic the solution) is, the lower the pH will be. Very acidic solutions have high hydrogen ion concentrations, low hydroxyl ion concentrations, low pH's, and high pOH's. In basic solutions the reverse is true (Fig. 3.1).

3–2. ACID-BASE EQUILIBRIUM PROBLEMS

There are primarily two types of simple acid-base problems: (a) those involving the acid or the base alone in water, which we shall call "Type 1" problems, and (b) those involving mixtures of acids and their conjugate bases, which we shall call "Type 2" problems.

Type 1 Problem: Acid Alone

A problem involving an acid alone in water was solved in Chapter 2 for the case of a 1.0 M solution of HF. Let us solve another problem of this type, including the K_w expression (3.16), and make the approximations necessary to obtain a simple solution. A table of K_a's is given in Appendix C.

Example 3.3. Calculate the $[H^+]$ in a solution prepared by diluting 0.10 mole of HCN to 1.00 liter with water. K_a for HCN is 7.2×10^{-10}.

(1)   Chemistry: $HCN \rightleftarrows H^+ + CN^-$ $H_2O \rightleftarrows H^+ + OH^-$

(2) $\dfrac{[H^+][CN^-]}{[HCN]} = 7.2 \times 10^{-10}$ $[H^+][OH^-] = 1.0 \times 10^{-14}$

(3) $$[H^+] = [CN^-] + [OH^-]$$

All problems involving acids and bases in water will include both H^+ and OH^- in the electroneutrality condition. Calculations are greatly simplified if one or the other of these can be neglected, depending upon whether we believe the solution to be acidic or basic. In this case, since only an acid has been added to water, the solution will be acidic, and we can try the approximation

$$[H^+] \gg [OH^-]$$

so that the electroneutrality condition becomes

$$[H^+] = [CN^-]$$

(4) $$[HCN] + [CN^-] = 0.10 \ M$$

Combining this material balance equation for cyanide with the electro-neutrality equation, we obtain

$$[HCN] + [H^+] = 0.10$$

Since K_a is small, most of the HCN will not be ionized, and therefore the concentration of undissociated HCN will be much larger than the $[H^+]$, and

$$[HCN] \gg [H^+] \quad \text{or} \quad [HCN] \approx 0.10 \ M$$

Using these equations and the equilibrium constant expression, we obtain

$$\frac{[H^+]^2}{0.10} = 7.2 \times 10^{-10}$$

$$[H^+] = 8.5 \times 10^{-6} \ M \quad \textit{Answer}$$

To test the approximations, we note that

$$[OH^-] = \frac{1.0 \times 10^{-14}}{8.5 \times 10^{-6}} = 1.2 \times 10^{-9} \ M$$

and the $[H^+]$ is indeed much larger than the $[OH^-]$. Similarly we find that the $[HCN]$ is much larger than the $[H^+]$. If the results of the calculation indicated that the approximations made did not hold, we would have to make new approximations, or rigorously solve the quadratic, cubic, or higher degree equations. Mathematical methods for solving problems without approximations are discussed in Chapter 10.

Type 1 Problem: Base Alone

The K_b for a base B is simply related to the K_a of its conjugate acid, HB^+ (or a base X^- and its conjugate acid HX). Base B ionizes according to the reaction

$$B + H_2O \rightleftarrows HB^+ + OH^- \tag{3.25}$$

and yields the equilibrium constant expression

$$\frac{[HB^+][OH^-]}{[B]} = K_b \tag{3.26}$$

Similarly, HB^+ ionizes according to the equation

$$HB^+ \rightleftarrows H^+ + B \tag{3.27}$$

and yields the equilibrium constant expression

$$\frac{[H^+][B]}{[HB^+]} = K_a \tag{3.28}$$

From (3.26) and (3.28) we obtain

$$K_aK_b = \frac{[H^+][B]}{[HB^+]} \frac{[HB^+][OH]}{[B]} = [H^+][OH^-] \tag{3.29}$$

or

$$K_aK_b = K_w \tag{3.30}$$

Knowing the K_a of an acid HB^+, we can easily calculate the K_b of its conjugate base B using (3.30). Appendix C contains no table of K_b's. To obtain the K_b of a given base, locate its conjugate acid, look up its K_a, and calculate the K_b with (3.30).

The calculation of the $[H^+]$ and $[OH^-]$ in solutions of a base alone in water is considered in the following example.

Example 3.4. Calculate the $[H^+]$ and $[OH^-]$ in a solution prepared by diluting 0.10 mole of NaCN to 1.0 liter with water.

(1) Chemistry: NaCN is completely ionized into Na^+ and CN^-. The base CN^- produces hydroxide ions by the reaction

$$CN^- + H_2O \rightleftarrows HCN + OH^-$$

(2)
$$\frac{[OH^-][HCN]}{[CN^-]} = K_b = 1.4 \times 10^{-5}$$

Since $K_a = 7.2 \times 10^{-10}$ for HCN,

$$K_b = \frac{K_w}{K_a} = \frac{1.0 \times 10^{-14}}{7.2 \times 10^{-10}} = 1.4 \times 10^{-5}$$

(3) $[H^+] + [Na^+] = [CN^-] + [OH^-]$

(4) $[Na^+] = 0.10\ M$ $[HCN] + [CN^-] = 0.10\ M$

Putting the $[Na^+]$ into the electroneutrality condition, and making the approximation that $[OH^-] \gg [H^+]$ (since the solution contains a base in water), we obtain

$$0.10 = [CN^-] + [OH^-]$$

Combining this equation with the material balance condition for cyanide gives,

$$[CN^-] + [OH^-] = 0.10 = [HCN] + [CN^-]$$
$$[OH^-] = [HCN]$$

Making the approximation that $[OH^-] \ll [CN^-]$, and substituting into the equilibrium constant expression we get

$$\frac{[OH^-]^2}{0.10} = 1.4 \times 10^{-5}$$

$$[OH^-] = 1.2 \times 10^{-3}\ M \quad \textit{Answer}$$

$$[H^+] = 8.5 \times 10^{-12}\ M \quad \textit{Answer}$$

Since $1.2 \times 10^{-3} \gg 8.5 \times 10^{-12}$ $([OH^-] \gg [H^+])$

and $0.10 \gg 1.2 \times 10^{-3}$ $([CN^-] \gg [OH^-])$

the approximations are valid.

Admittedly, it is often difficult to know in advance whether an approximation can or cannot be made. It is always permissible to try an approximation and then examine the results to see if the assumptions are confirmed. The other alternative is to make no approximations and to solve the more difficult algebraic expressions. For example, if the assumption that the $[CN^-]$ is much larger than the $[OH^-]$ is not made, substitution in the equilibrium expression yields

$$\frac{[OH^-]^2}{0.10 - [OH^-]} = K_b = 1.4 \times 10^{-5} \tag{3.31}$$

so that the equation that must be solved is

$$[OH^-]^2 + K_b[OH^-] - 0.10\ K_b = 0 \tag{3.32}$$

If no assumptions are made, the $[H^+]$ is included in the electroneutrality condition and the K_w expression must be used, the equation that results is

$$[OH^-]^3 + K_b[OH^-]^2 + 0.10\ K_b\ [OH^-] - K_w K_b = 0 \tag{3.33}$$

Solution of either (3.32) or (3.33) yields (with some mathematical labor) the same value for the $[OH^-]$ as obtained by use of approximations. The reader will appreciate the value of making logical approximations and greatly simplifying the labors of solving difficult algebraic equations.

Type 2 Problems: Acid-Base Mixtures

A problem involving a weak acid and its conjugate base is solved by the same procedure as before.

Example 3.5.   Calculate the $[H^+]$ in a solution prepared by diluting 0.10 mole of HCN and 0.10 mole of NaCN to 1.0 liter with water.

(1) $$HCN \rightleftarrows H^+ + CN^-$$

(2) $$\frac{[H^+][CN^-]}{[HCN]} = K_a = 7.2 \times 10^{-10}$$

(3) $$[Na^+] + [H^+] = [OH^-] + [CN^-]$$

(4) $$[Na^+] = 0.10 \ M \quad [HCN] + [CN^-] = 0.20 \ M$$

From the electroneutrality condition and Na^+ material balance we obtain

$$[CN^-] = 0.10 + [H^+] - [OH^-]$$

Making the approximation that $[H^+] - [OH^-] \ll 0.10$ (or $[H^+] \ll 0.10$ and $[OH^-] \ll [CN^-]$) we find $[CN^-] \approx 0.10$ and $[HCN] \approx 0.20 - 0.10 \approx 0.10$. Introducing these approximations into the equilibrium constant expression we further obtain

$$\frac{[H^+](0.10)}{(0.10)} = 7.2 \times 10^{-10} \ M = [H^+] \quad Answer$$

$$[OH^-] = \frac{1.0 \times 10^{-14}}{7.2 \times 10^{-10}} = 1.4 \times 10^{-5} \ M \quad Answer$$

Note that the $[H^+]$ found in Example 3.5 is much smaller than that in Example 3.3; i.e., that the presence of additional cyanide ion lowers the $[H^+]$ by a factor of almost 10,000. This effect is an example of Le Chatelier's principle; addition of CN^- shifts the equilibrium

$$HCN \rightleftarrows H^+ + CN^-$$

to the left and decreases the $[H^+]$. This is called the *common ion effect*.

Most simple acid-base problems reduce to either a Type 1 or a Type 2 problem. For example, we can obtain the same conditions as in Example 3.3 (a solution containing only 0.10 mole of HCN per liter of solution) by mixing 0.10 mole of NaCN and 0.10 mole of HCl and diluting to 1.0 liter with water. We must realize that a *neutralization* reaction occurs when an acid (HCl) and a base (CN^-) are mixed

$$H^+ + CN^- \rightleftarrows HCN \qquad (3.34)$$

Since the equilibrium constant for this reaction as written is large (equal to $1/K_a$ for HCN ionization), essentially 0.10 mole of HCN will be produced, and nearly all of the H^+ and CN^- will be used up. After this initial consideration, the problem is solved as a typical Type 1 problem. It could also be solved by setting up the electroneutrality condition and material balances on the original mixture (HCl + NaCN).

Preliminary neutralization reactions may also lead to Type 2 problems. For example a mixture of 0.10 M HCN and 0.10 M CN$^-$, as in Example 3.5, can also be obtained by two different neutralizations.

(1) Mix 0.10 mole of HCl and 0.20 mole of NaCN and dilute to 1.0 liter. Again the reaction of H$^+$ and CN$^-$ occurs and 0.10 mole of H$^+$ reacts with 0.10 mole of CN$^-$, this time leaving 0.10 mole of CN$^-$ excess in solution.

(2) Mix 0.20 mole of HCN with 0.10 mole of NaOH and dilute to 1.0 liter. This time the neutralization reaction is

$$\text{HCN} + \text{OH}^- \rightleftharpoons \text{CN}^- + \text{H}_2\text{O} \qquad (3.35)$$

Here 0.10 mole of OH$^-$ neutralizes essentially 0.10 mole of HCN, leaving a solution containing 0.10 mole of HCN and 0.10 mole of CN$^-$. Again, after these preliminary considerations, these problems become almost identical with Example 3.5, and can be solved by the method described there.

Example 3.6. Calculate the [OH$^-$] in a solution prepared by mixing 1.0 mole of HCl, 1.0 mole of Ba(OH)$_2$, and 1.5 moles of NH$_4$Cl, and diluting this mixture to 1.0 liter.

(1) $\quad$ NH$_4{}^+$ + OH$^-$ $\rightleftharpoons$ NH$_3$ + H$_2$O $\quad$ H$^+$ + OH$^-$ $\rightleftharpoons$ H$_2$O

(2) $\quad \dfrac{[\text{NH}_4{}^+][\text{OH}^-]}{[\text{NH}_3]} = K_b = 1.8 \times 10^{-5} \quad [\text{H}^+][\text{OH}^-] = K_w$

(3) $\quad [\text{H}^+] + 2[\text{Ba}^{++}] + [\text{NH}_4{}^+] = [\text{Cl}^-] + [\text{OH}^-]$

(4) $\quad [\text{Ba}^{++}] = 1.0 \ M \quad [\text{Cl}^-] = 2.5 \ M$

$$[\text{NH}_4{}^+] + [\text{NH}_3] = 1.5$$

Put the concentrations of Ba^{++} and Cl$^-$ into the electroneutrality condition

$$[\text{H}^+] + 2(1.0) + [\text{NH}_4{}^+] = 2.5 + [\text{OH}^-]$$

With the assumption that

$$[\text{H}^+] \ll 2.0 \quad \text{and} \quad [\text{OH}^-] \ll 2.5$$

$$[\text{NH}_4{}^+] \approx 0.5 \ M$$

$$[\text{NH}_3] = 1.5 - 0.5 \approx 1.0 \ M$$

and, using the equilibrium constant expression for the basicity of NH$_3$

(or for the acidity of NH_4^+) we can solve for $[OH^-]$ and $[H^+]$

$$\frac{(0.5)[OH^-]}{(1.0)} = 1.8 \times 10^{-5}$$

$$[OH^-] = 3.6 \times 10^{-5} \, M \quad \textit{Answer}$$

$$[H^+] = \frac{1.0 \times 10^{-14}}{3.6 \times 10^{-5}} = 2.8 \times 10^{-10} \, M$$

This problem can also be solved using the following reasoning. The 1.0 mole of H^+ (from the HCl) will neutralize 1.0 of the 2.0 moles of OH^- (from the $Ba(OH)_2$) leaving 1.0 mole of OH^- in excess. This will react with 1.0 of the 1.5 moles of NH_4^+ and will finally produce a solution containing essentially 0.5 mole of NH_4^+ and 1.0 mole of NH_3. These concentrations (the same as those obtained by explicit approximations in the above method) are then used in the K_b expression as before. This approach is much faster and is frequently useful for simple systems. It requires a better understanding of the chemistry of the situation than does the more rigorous solution of the problem, and is not as straightforward when the problem contains some complications (for example if the problem involved acetic acid rather than HCl).

3–3. POLYPROTIC ACIDS

So far only acids containing a single proton, *monoprotic acids*, have been considered. There are many acids which have several protons (*polyprotic acids*), such as the diprotic acids H_2S, H_2CO_3, and H_2SO_4, and the triprotic acids H_3PO_4 and H_3AsO_4. Although equilibrium problems involving polyprotic acids are solved by the same methods as those developed for monoprotic acids, they are usually more complex because of the additional equilibria involved. Ionization of polyprotic acids occurs stepwise; each ionization step involves an equilibrium constant expression. The acid H_2X ionizes according to the following equations

$$H_2X \rightleftarrows H^+ + HX^- \tag{3.36}$$

$$\frac{[H^+][HX^-]}{[H_2X]} = K_1 \tag{3.37}$$

$$HX^- \rightleftarrows H^+ + X^{--} \tag{3.38}$$

$$\frac{[H^+][X^{--}]}{[HX^-]} = K_2 \tag{3.39}$$

Example 3.7. Calculate the $[H^+]$, $[HS^-]$, and $[S^{--}]$ in a solution prepared by dissolving 0.10 mole of H_2S in sufficient water to produce 1.0 liter of solution. For H_2S, $K_1 = 1.1 \times 10^{-7}$, $K_2 = 1 \times 10^{-14}$.

(1)
$$H_2S \rightleftarrows H^+ + HS^- \qquad HS^- \rightleftarrows H^+ + S^{--}$$

(2)
$$\frac{[H^+][HS^-]}{[H_2S]} = 1.1 \times 10^{-7} \qquad \frac{[H^+][S^{--}]}{[HS^-]} = 1 \times 10^{-14}$$

(3)
$$[H^+] = [HS^-] + 2[S^{--}]$$

(assuming that the solution is acidic; i.e., that $[H^+] \gg [OH^-]$)

(4)
$$[H_2S] + [HS^-] + [S^{--}] = 0.10$$

The reader may confirm that solving the above equations without any approximations results in a cubic equation in $[H^+]$ (generally a problem involving n K_a and K_b expressions leads to an $n + 1$ degree equation in $[H^+]$). Since K_2 is very small, only a very small amount of HS^- is ionized to produce S^{--}, and the approximation

$$[HS^-] \gg [S^{--}]$$

can be made, so that the electroneutrality condition becomes

$$[H^+] \approx [HS^-]$$

Since K_1 is also quite small

$$[H_2S] \gg [HS^-]$$

and from the material balance equation

$$[H_2S] \approx 0.10 \ M$$

With these approximations introduced into the K_1 expression, we obtain

$$\frac{[H^+]^2}{0.10} = 1.1 \times 10^{-7}$$

$$[H^+] = [HS^-] = 1.0 \times 10^{-4} \ M \quad Answer$$

The $[S^{--}]$ is determined from the K_2 expression

$$\frac{[H^+][S^{--}]}{[HS^-]} = \frac{(1.0 \times 10^{-4})[S^{--}]}{(1.0 \times 10^{-4})} = 1 \times 10^{-14}$$

$$[S^{--}] = 1 \times 10^{-14} \ M \quad Answer$$

Also

$$[OH^-] = \frac{1.0 \times 10^{-14}}{1.0 \times 10^{-4}} = 1.0 \times 10^{-10} \ M$$

Note that the results obtained in this case justify the approximations.

Solution of Type 2 polyprotic acid problems involves similar considerations.

Example 3.8. Calculate the $[H^+]$, $[HS^-]$, and $[S^{--}]$ in a solution prepared by dissolving 0.10 mole of H_2S and 0.10 mole of HCl in water to produce 1.0 liter of solution.

The (1) ionization reactions and (2) equilibrium constant expressions are the same as those given in Example 3.7.

(3) $$[H^+] = [HS^-] + 2[S^{--}] + [Cl^-]$$

(4) $$[H_2S] + [HS^-] + [S^{--}] = 0.10 \quad [Cl^-] = 0.10 \ M$$

Putting $[Cl^-]$ into the electroneutrality condition yields

$$[H^+] = [HS^-] + 2[S^{--}] + 0.10$$

or, with the assumption: $[HS^-] + 2[S^{--}] \ll 0.10$,

$$[H^+] \approx 0.10 \ M$$

As before

$$[H_2S] \approx 0.10 \ M$$

Applying these approximations to the K_1 expression

$$\frac{(0.10)[HS^-]}{(0.10)} = 1.1 \times 10^{-7}$$

$$[HS^-] = 1.1 \times 10^{-7} \ M \quad \textit{Answer}$$

From the K_2 expression

$$\frac{(0.10)[S^{--}]}{(1.1 \times 10^{-7})} = 1.0 \times 10^{-14}$$

$$[S^{--}] = 1.1 \times 10^{-20} \ M \quad \textit{Answer}$$

Notice that the addition of a common ion again shifts the equilibria to the left and decreases the concentrations of HS^- and S^{--}, in this case as compared to those in Example 3.7.

For triprotic acids the rigorous general solution (including the K_w expression) results in a fifth-degree equation in $[H^+]$. Usually, however, approximations of the type employed in Examples 3.7 and 3.8 lead to considerable simplification. These problems become more complicated when (1) K_1 and K_2 are numerically about the same (e.g., for succinic acid, $K_1 = 6.3 \times 10^{-5}$ and $K_2 = 3.3 \times 10^{-6}$), so that the concentrations of HX^- and X^{--} are of comparable magnitude, or (2) when K_1 is large (e.g., for H_3PO_4, $K_1 = 6 \times 10^{-3}$), so that an appreciable amount of the parent acid is ionized. In both of these cases the rigorous algebraic equations can be solved using well known, but unfortunately tedious, mathematical methods (Chapter 10). Problems involving mixtures of weak acids and bases may also be quite complicated and result in algebraic equations of degree greater than 2. For example the reader might consider the problem of finding the pH of a solution containing 0.10 mole of $(NH_4)_2HPO_4$ per liter. A means of solving complex problems of this type, based upon graphical methods, is discussed in Chapter 8.

3–4. NONAQUEOUS SOLUTIONS

Although our interest has been confined to aqueous solutions, a similar treatment of acids and bases in such solvents as glacial acetic acid, liquid ammonia, acetonitrile (CH_3CN), and others can be given. The discussion here will be restricted to solvents containing ionizable protons. As before an acid (HA) reacts with the solvent (HS) to form a solvated proton

$$HA + HS \rightleftarrows H_2S^+ + A^- \qquad (3.40)$$

Just as the reaction of HCl with water is given by

$$HCl + H_2O \rightarrow H_3O^+ + Cl^- \qquad (3.41)$$

so the reaction with glacial acetic acid (written HAc) is

$$HCl + HAc \rightleftarrows H_2Ac^+ + Cl^- \qquad (3.42)$$

and the reaction in liquid ammonia is

$$HCl + NH_3 \rightarrow NH_4^+ + Cl^- \qquad (3.43)$$

Since these reactions depend upon both the proton-donating ability of the acid, HA, and the proton-accepting ability of the solvent, HS, the "strength" of an acid, measured by the extent to which the ionization reaction (3.40) lies to the right, is strongly dependent upon the solvent itself. Water is a fairly good proton acceptor, and (3.41) lies completely to the right, so that HCl is a strong acid in aqueous solution. Glacial acetic acid on the other hand is a much poorer proton acceptor (a much less basic solvent), and (3.42) lies only partially to the right, so that HCl is a weak acid in this solvent. Ammonia, however, is a much better proton acceptor (a more basic solvent) than water, and not only HCl, but many acids that are weak in aqueous solution appear as strong acids in liquid ammonia. A solvent which is a good proton acceptor is said to exert a *leveling effect* upon acids, so that many acids appear as strong acids in it, and the inherent strength of the individual acids cannot be discerned. For example, in water, $HClO_4$, HNO_3, and HCl are all essentially completely ionized and the relative strengths of these three acids cannot be determined. In glacial acetic acid however, $HClO_4$ is found to be a stronger acid than either HNO_3 or HCl. Liquid ammonia exerts an even stronger leveling effect than water on acids, and even HF is a strong acid in liquid ammonia.

The behavior of bases in nonaqueous solvent follows in an analogous manner. A base, B, accepts a proton from the solvent, HS

$$B + HS \rightleftarrows HB^+ + S^- \tag{3.44}$$

and the strength of a base depends upon the extent to which (3.44) lies to the right. Bases that are weak in water will appear stronger in a solvent with greater proton donating ability (a more acidic solvent), such as glacial acetic acid.

A nonaqueous solvent may undergo self-ionization (*autoprotolysis*) just as water does, by the reaction

$$HS + HS \rightleftarrows H_2S^+ + S^- \tag{3.45}$$

yielding the *autoprotolysis constant* expression (analogous to (3.16) for water)

$$[H_2S^+][S^-] = K_s \tag{3.46}$$

The K_a and K_b of a conjugate acid-base pair in a nonaqueous solvent are related by the expression

$$K_a K_b = K_s \tag{3.47}$$

which is a generalized form of (3.30).

Problems in nonaqueous solvents are sometimes complicated by *ion pair formation*. In solvents of low dielectric constant (essentially low electrical insulating ability), oppositely charged ions are attracted to one another and a large fraction of the ions in solution may be present in the form of these clusters or ion pairs. This behavior necessitates consideration of additional equilibria of the type

$$H_2S^+ A^- \rightleftarrows H_2S^+ + A^- \tag{3.48}$$

Since water has a high dielectric constant, ion pair formation is not very important in aqueous solutions. In a solvent such as acetic acid however, in addition to the usual acid-base equilibria, ion pair formation must be taken into account.

Nonaqueous solutions have proven to be a fruitful field of chemical research. Interest in this field is motivated by the application of these solvents to the study of reactions which cannot be carried out in water (such as the reaction of an extremely weak base with acid in glacial acetic acid, or reactions involving acids or bases insoluble in water), by the existence of nonaqueous seas on other planets (such as liquid ammonia on Jupiter), and because a study of nonaqueous systems gives better insight into the behavior of acids and bases in water. Some solvents, such as glacial acetic acid and liquid ammonia have been studied extensively, but many others have not been investigated in any quantitative detail.

PROBLEMS

3.1. Calculate the pH, pOH, and [OH$^-$] in solutions of the following hydrogen ion concentrations.

(a) $2.0 \times 10^{-15}\ M$	(g) $8.0 \times 10^{-4}\ M$
(b) $3.0 \times 10^{-12}\ M$	(h) $9.0 \times 10^{-3}\ M$
(c) $4.0 \times 10^{-10}\ M$	(i) $0.20\ M$
(d) $5.0 \times 10^{-8}\ M$	(j) $1.0\ M$
(e) $6.0 \times 10^{-7}\ M$	(k) $5.0\ M$
(f) $7.0 \times 10^{-5}\ M$	(l) $15\ M$

3.2. Calculate the pOH, [OH$^-$], and [H$^+$] in solutions with the following pH's.

(a) -1.10	(d) 1.10	(g) 7.52	(j) 12.60
(b) -0.56	(e) 3.30	(h) 8.80	(k) 14.00
(c) 0.0	(f) 5.60	(i) 10.90	(l) 14.70

3.3. Calculate the hydrogen ion concentration and the pH of each of the following solutions.

(a) 0.0020 moles of HCl diluted to 500 ml

(b) 0.15 gram HNO_3 diluted to 300 ml

(c) 10.0 ml of $15\ M$ HCl diluted to 750 ml

(d) 0.050 mole NaOH diluted to 400 ml

(e) 2.5 grams KOH diluted to 750 ml

(f) 1.0×10^{-8} mole HCl diluted to 1.0 liter (Hint: This problem must be solved rigorously, writing ionic and material balances if a reasonable answer is to be obtained.)

(g) 1.0×10^{-7} moles NaOH diluted to 1.0 liter (see hint in f.)

(h) 7.46 grams KCl diluted to 500 ml

(i) 0.10 mole $NaNO_3$ diluted to 1.0 liter

3.4. Calculate the concentrations of all of the species, ions and molecules, and the pH, of the following solutions. All solutions have a total volume of 1.00 liter and contain:

(a) 0.10 mole HClO	(h) 0.10 mole H_2Se
(b) 0.40 mole HNO_2	(i) 0.20 mole H_2CO_3
(c) 0.050 mole HCOOH	(j) 0.10 mole NaClO
(d) 0.10 mole C_6H_5OH	(k) 0.10 mole KNO_2
(e) 0.20 mole NH_3	(l) 0.10 mole NH_4Cl
(f) 0.10 mole C_5H_5N	(m) 0.20 mole C_5H_5NHCl
(g) 0.10 mole H_2SO_3	

3.5. Calculate the concentrations of all of the species, ions and molecules, and the pH, of the following solutions. All solutions have a total volume of 1.00 liter and contain:

(a) 0.20 mole NaClO, 0.10 mole HClO

(b) 1.0 mole KNO_2, 0.20 mole HNO_2

(c) 0.10 mole NaCOOH, 0.40 mole HCOOH

(d) 0.10 mole $NaOC_6H_5$, 1.0 mole HOC_6H_5

(e) 0.20 mole NH_3, 0.50 mole NH_4Cl

(f) 0.10 mole $C_6H_5NH_2$, 0.40 mole $C_6H_5NH_3Cl$

(g) 0.20 mole HCl, 0.10 mole H_2S

(h) 1.0 mole HNO_3, 0.10 mole H_2Se

(i) 0.0010 mole HCl, 0.010 mole H_2CO_3

3.6. Calculate the concentrations of all of the species, ions and molecules, and the pH, of the following solutions. All solutions have a total volume of 1.00 liter and contain:

(a) 0.50 mole HCl, 0.20 mole NaOH, and 1.00 mole CH_3COONa

(b) 2.00 mole HCl, 0.20 mole NaOH, and 1.00 mole CH_3COONa

(c) 1.00 mole NH_3, 1.00 mole NH_4Cl, and 0.40 mole NaOH

(d) 2.00 mole NH_3, 1.00 mole HCl

(e) 1.00 mole NaAc, 0.50 mole HCl

3.7. An *acid-base titration* involves the addition of increments of an acid solution to a base solution (or vice-versa).

(a) A chemist adds 50.0 ml of a 0.10 M NaOH solution to a flask and adds various volumes of 0.10 M HCl solution. Calculate the pH of the resulting mixture in the flask after the addition of 0, 10.0, 40.0, 50.0, 51.0, and 60.0 ml of the HCl solution.

(b) Draw a graph using volume of HCl taken as the x-axis and pH as the y-axis. Plot the points calculated in (a) and sketch the *titration curve*.

3.8. Repeat the calculations and graph described in Problem 3.7 for the following systems:

(a) 0.10 M HCl titrated with 0.10 M NaOH
(b) 0.10 M CH_3COOH titrated with 0.10 M NaOH
(c) 0.10 M NH_3 titrated with 0.10 M HCl
(d) 0.10 M HCN titrated with 0.10 M NaOH

3.9. How many grams of sodium benzoate, C_6H_5COONa, must be added to 250 ml of 0.010 M C_6H_5COOH solution to produce a solution of pH $= 4.80$?

3.10. When 0.10 mole of acid HX is diluted to 1.00 liter, a solution of pH $= 2.60$ results. What is the K_a of HX?

3.11. A *buffer* solution is one that resists changes in pH and is usually composed of a conjugate acid-base mixture. To illustrate the action of a buffer, consider 1.00 liter of a solution containing 0.10 mole of CH_3COOH and 0.10 mole CH_3COONa. (a) Calculate the pH of this buffer solution. (b) Calculate the pH after the addition of 0.010 mole NaOH to the original buffer solution.

To contrast the action of a buffered solution to that of an unbuffered one: (c) calculate the number of moles of HCl that must be added to 1 liter of water to produce a solution with a pH of that in part (a). (d) Calculate the pH of this *unbuffered* solution after the addition of 0.010 moles of HCl. (e) Calculate the pH after the addition of 0.010 moles of NaOH to the original unbuffered solution.

3.12. Calculate the ratio of CH_3COOH to CH_3COO^- in solutions having pH of (a) 2.00 (b) 3.00 (c) 5.00 (d) 7.00 (e) 9.00.

3.13. In dilute solutions of very weak acids, the $[OH^-]$ cannot be considered small compared to the other species, and the water equilibrium expression must be included in the various equations used in solving the problem in order to obtain a reasonable answer. Calculate the concentration of all of the species, ions and molecules, and the pH of a solution containing 1.0×10^{-4} mole of HBO_2 in 1 liter of solution.

3.14. Write the K_a and K_b expressions for the ionization of HX and X^- respectively in a general nonaqueous solvent, HS. Derive Equation (3.47).

SUPPLEMENTARY READING

Acids and Bases, *J. Chem. Educ. Publ.* No. 1, 1941. An interesting collection of papers discussing the relative merits of different acid-base definitions, acidity and basicity in nonprotonic solvents, and other topics.

Davidson, D., Amphoteric Molecules and Salts, *J. Chem. Educ.*, **32**, 550 (1955).

DeFord, D., The Brønsted Concept in Acid-Base Calculations, *J. Chem. Educ.*, **27**, 554 (1950).

King, E. J., *Acid-Base Equilibria*, New York: Macmillan, 1965.

Kolthoff, I. M., and S. Bruckenstein, in I. M. Kolthoff and P. J. Elving, *Treatise on Analytical Chemistry*, New York: Interscience, 1959, chps. 11, 12, and 13.

Ricci, J., *Hydrogen Ion Concentration*, Princeton, N.J.: Princeton University Press, 1952. A complete treatment of acid-base calculations, slightly difficult to read because of the unconventional notation adopted.

CHAPTER 4 SOLUBILITY EQUILIBRIA

4–1. THE SOLUBILITY PRODUCT

SOLUBILITY EQUILIBRIA

Equilibrium theory is frequently applied to problems involving slightly soluble electrolytes. Suppose we add several grams of silver chloride, AgCl, to water. Although AgCl is not very soluble, some small amount will dissolve to produce silver ions and chloride ions*

$$AgCl(solid) \rightleftarrows Ag^+ + Cl^- \qquad (4.1)$$

As soon as some AgCl dissolves, a back-reaction, the reprecipitation of Ag^+ and Cl^-, begins. Eventually the rate of dissolution and the rate of precipitation become equal and a dynamic equilibrium is established. Unlike acid-base reactions, which are usually very fast,

---

* In this initial discussion of solubility equilibria we will neglect such complicating effects as complex ion formation, undissociated AgCl in solution, and other competing equilibria for the sake of simplicity. These factors will be considered later.

solubility reactions may take a long time to attain equilibrium. A solid in contact with a solution can be shown to be a condition of dynamic equilibrium, rather than in a static condition, by considering the fate of an impurity ion trapped inside a solid crystal lattice of a slightly soluble compound. With time, because of the continuous dissolution and reprecipitation, the impurity may find itself on the surface of the crystal and may eventually dissolve and be identified in the solution. However, if no true equilibrium existed, and some ions merely dissolved, and all reactions stopped, such deep-seated impurities would never reach the surface and dissolve.

The dynamic nature of solubility equilibria can also be demonstrated with the aid of radioactive tracers. For example, suppose some solid AgI is put into a test tube and a solution containing radioactive iodide ion (iodine-131) were poured over it (Fig. 4.1). Initially the precipitate, containing only stable (nonradioactive) iodide, shows no activity on a Geiger Counter, while the solution, containing radioactive iodide (I^-*) shows high radioactivity. Since the dynamic equilibrium

$$AgI(solid) \rightleftarrows Ag^+ + I^- \tag{4.2}$$

occurs, with time radioactive iodide is incorporated into the precipitate and stable iodide ion is liberated into the solution. The radioactivity of the precipitate increases while that of the solution decreases. By measuring the radioactivity of the precipitate and the solution at various times, the rate at which the system, which can also be represented as

$$AgI(solid) + I^-* \rightleftarrows AgI*(solid) + I^- \tag{4.3}$$

attains equilibrium, can be determined.

THE K_{sp} EXPRESSION

Once we are convinced that (4.1) is an equilibrium reaction, we can apply our previously developed theory, and write the equilibrium constant expression

$$\frac{[Ag^+][Cl^-]}{AgCl(solid)} = K \tag{4.4}$$

At first glance (4.4) appears troublesome. Although the notion of $[Ag^+]$ and $[Cl^-]$ is clear, what will we use for the term AgCl(solid)? The following argument suggests that AgCl (solid) is a constant.

Let us imagine the dissolution of sodium chloride, NaCl, in water. The process here is exactly the same as the one for AgCl, but since

Initially

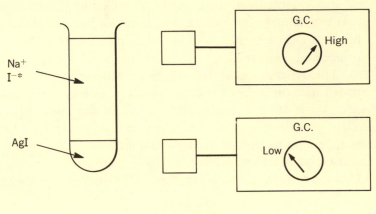

$$AgI + I^{-}* \leftrightarrows AgI^* + I^-$$

After equilibrium is attained

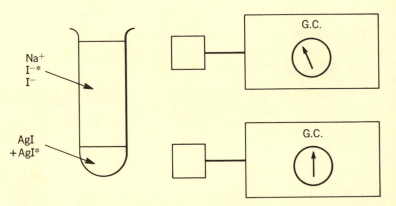

FIGURE 4.1. A demonstration of solubility equilibrium. Initially all of the radioactive iodide (I⁻*) is in the solution, so that the Geiger Counter (G. C.) shows a high solution activity, but only a small amount of radioactivity in the AgI precipitate. After equilibrium is attained, radioactivity is found in both solution and precipitate.

NaCl is so much more soluble, the suggestion of a solubility limit is easier to imagine. As we add NaCl to water it will dissolve until a point is reached when we drop a small crystal of NaCl into the solution and it does not dissolve. At this point we say we have produced a *saturated solution* of NaCl. All the NaCl that can dissolve at this particular temperature has dissolved. No matter how much additional NaCl we now add to the beaker, be it a spoonful, a pound, or even a ton, the limit of solubility has been reached, and the concentrations of sodium and chloride ions in the solution are constant. Going back now to the AgCl case, we see that as long as we consider saturated solutions, the term, AgCl(solid), is a constant. Under these conditions we can combine it with the equilibrium constant, K, and obtain a new constant, K_{sp}, the *solubility product constant*.

$$[Ag^+][Cl^-] = K(AgCl(solid)) = K_{sp} \qquad (4.5)$$

(4.5) indicates that in a saturated solution of AgCl, i.e., a solution in equilibrium with at least a tiny crystal of solid AgCl, the product of the concentrations of silver ion and chloride ion, no matter what their source, is a constant.

For a general, slightly soluble, compound, M_mN_n, the dissolution equation is

$$M_mN_n(solid) \rightleftarrows mM^{+n} + nN^{-m} \qquad (4.6)$$

and the solubility product expression is

$$[M^{+n}]^m[N^{-m}]^n = K_{sp} \qquad (4.7)$$

Remember that the solubility product expression, unlike most other equilibrium constant expressions, contains no denominator.

4–2. SOLUBILITY PROBLEMS

Just as with acid-base problems, we can discuss
(a) Type 1 problems, involving the precipitate or solid alone, with no common ions in the solution, and
(b) Type 2 problems, where another source of ions contained in the precipitate is present in the solution.

Type 1 Problems

When a slightly soluble electrolyte is mixed with water, and a saturated solution is formed, we can calculate the solubility of the precipitate and ionic concentrations in the solution, using the K_{sp} expression.

Example 4.1. Calculate the $[Ag^+]$ and $[CrO_4^{--}]$, and the solubility of Ag_2CrO_4 in mg/liter, in a solution prepared by diluting 10.0 grams of Ag_2CrO_4 to 250 ml with water.

(1) $$Ag_2CrO_4(\text{solid}) \rightleftarrows 2Ag^+ + CrO_4^{--} \tag{4.8}$$

(2) $$[Ag^+]^2\,[CrO_4^{--}] = K_{sp} = 1.9 \times 10^{-12} \tag{4.9}$$

A table of K_{sp}'s of a number of slightly soluble compounds is given in Appendix C.

(3) $$[Ag^+] = 2[CrO_4^{--}] \tag{4.10}$$

The electroneutrality condition in this case involves only Ag^+ and CrO_4^{--}. It is not necessary to consider the water equilibrium in this problem, because neither H^+ nor OH^- enter into the above equilibrium under these conditions.

(4) The preceding two equations supply sufficient conditions to solve for the two unknowns. (4.10) is also the material balance equation, since for every mole of Ag_2CrO_4 that dissolves, two moles of Ag^+ and one mole of CrO_4^{--} are produced. (4.9) and (4.10) can be combined to yield

$$[Ag^+]^2 = 4[CrO_4^{--}]^2$$

$$4[CrO_4^{--}]^3 = 1.9 \times 10^{-12}$$

$$[CrO_4^{--}] = 0.78 \times 10^{-4}\,M \quad \textit{Answer}$$

$$[Ag^+] = 2[CrO_4^{--}] = 1.56 \times 10^{-4}\,M \quad \textit{Answer}$$

The solubility is the number of moles of solid which dissolves per liter of solution. Since every time one mole of silver chromate dissolves, one mole of CrO_4^{--} is produced, the solubility in this case is 0.78×10^{-4} moles/liter. The solubility can be expressed in grams/liter by multiplying by the molecular weight of Ag_2CrO_4

$$\text{solubility} = 0.78 \times 10^{-4}\ \text{moles/liter} \times 332\ \text{grams/mole}$$

$$\text{solubility} = 0.0258\ \text{grams/liter} = 25.8\ \text{mg/liter}$$

Notice that the fact that 10.0 grams of Ag_2CrO_4 was taken, or that the volume of the solution was 250 ml, was not used in solving the problem. As long as an excess of solid is present, so that the solution is saturated, the solubility is not affected by how large that excess is. Therefore as long as more than 25.8 mg of Ag_2CrO_4 are added per liter of solution, the results will be the same. It is also important to differentiate between the *solubility* of silver chromate and the *solubility product* of silver chromate. The solubility simply indicates how many moles of solid Ag_2CrO_4 will dissolve in a liter of water, while the solubility product is a constant related to the Ag^+ and CrO_4^{--} concentrations in the solution.

Example 4.2. Calculate the $[Ce^{+++}]$ and $[IO_3^-]$ in a saturated solution of $Ce(IO_3)_3$ ($K_{sp} = 3.2 \times 10^{-10}$).

$$(1) \qquad Ce(IO_3)_3(solid) \rightleftarrows Ce^{+++} + 3IO_3^-$$

$$(2) \qquad [Ce^{+++}][IO_3^-]^3 = 3.2 \times 10^{-10}$$

$$(3) \qquad 3[Ce^{+++}] = [IO_3^-]$$

The final solution of this problem, found by combining the above two equations, is left to the reader.

Type 2 Problems

These problems involve precipitates in the presence of ions in solution in common with those in the precipitate, from sources other than the precipitate.

Example 4.3. Calculate the $[Ag^+]$ and $[CrO_4^{--}]$, and the solubility of Ag_2CrO_4 in a solution prepared by adding an excess of solid Ag_2CrO_4 and 0.10 mole of Na_2CrO_4 to a beaker and diluting with water to 1 liter.

$$(1) \qquad Ag_2CrO_4(solid) \rightleftarrows 2Ag^+ + CrO_4^{--}$$

$$(2) \qquad [Ag^+]^2[CrO_4^{--}] = 1.9 \times 10^{-12}$$

$$(3) \qquad [Ag^+] + [Na^+] = 2[CrO_4^{--}]$$

$$(4) \qquad [Na^+] = 0.20\ M$$

We now have three equations and three unknowns. From the last two equations we obtain

$$[CrO_4^{--}] = \tfrac{1}{2}[Ag^+] + 0.10$$

This equation combined with the K_{sp} expression yields a cubic equation in $[Ag^+]$. The mathematics is simpler if we make the assumption

$$[Ag^+] \ll 0.10\ M$$

so that

$$[CrO_4^{--}] \approx 0.10\ M$$

The solution of the equation with this assumption is

$$[Ag^+] = \sqrt{19 \times 10^{-12}} = 4.4 \times 10^{-6}\ M \quad Answer$$

The solubility of Ag_2CrO_4, equal to one-half the $[Ag^+]$, is 2.2×10^{-6} moles/liter. Notice the *common ion effect:* The solubility of Ag_2CrO_4 in this case is smaller than in Example 4.1; that is, the solubility is smaller in the presence of an excess of a common ion.

A useful variation of the Type 2 problem is one which allows the calculation of conditions necessary for precipitation to occur, or the calculation of the completeness of precipitation. These calculations are based upon the following principles.

(1) Precipitation will occur when the product of the concentrations of the ions, raised to the appropriate power in the K_{sp} expression, is larger than the K_{sp}.

(2) Given the concentration of any one ion in the K_{sp} expression, the concentration of the other ion can be calculated directly.

Example 4.4. What $[Ba^{++}]$ is necessary to start $BaSO_4$ precipitating from a solution which is $0.0010 \, M$ in SO_4^{--}?

The solubility product expression for $BaSO_4$ is

$$[Ba^{++}][SO_4^{--}] = 1.0 \times 10^{-10} \tag{4.11}$$

In a saturated solution of $BaSO_4$, the left-hand side of this equation (the *ion concentration product*) just equals the right-hand side. In an unsaturated solution, the ion concentration product is less than the K_{sp}, therefore, in a solution containing $0.0010 \, M \, SO_4^{--}$, no precipitation of $BaSO_4$ will occur, as long as

$$[Ba^{++}] < \frac{1.0 \times 10^{-10}}{0.0010} = 1.0 \times 10^{-7} \, M$$

Once the $[Ba^{++}]$ gets above $1.0 \times 10^{-7} \, M$, precipitation will occur and will continue until the K_{sp} expression, (4.11), is satisfied. Therefore, the $[Ba^{++}]$ for incipient precipitation is $1.0 \times 10^{-7} \, M$.

Example 4.5. When IO_3^- is added to a solution containing $0.10 \, M$ Ba^{++}, $Ba(IO_3)_2$ precipitates.

(a) What final concentration of IO_3^- is required to precipitate the Ba^{++} quantitatively (i.e., allow only 0.1 percent of the original Ba^{++} to remain in solution)?

If the $[Ba^{++}]$ is to be decreased to 0.1 percent of its initial amount, then the final barium ion concentration is given by

$$[Ba^{++}] = 10^{-3} \times 0.10 = 1.0 \times 10^{-4} \, M$$

$$[Ba^{++}][IO_3^-]^2 = 1.5 \times 10^{-9}$$

To produce this $[Ba^{++}]$, the $[IO_3^-]$ must be

$$[IO_3^-] = \sqrt{\frac{1.5 \times 10^{-9}}{1.0 \times 10^{-4}}} = 3.9 \times 10^{-3} \, M \quad Answer \quad (4.12)$$

(b) How many moles of KIO_3 must be added per liter of solution to bring about this condition?

For 1 liter of solution, initially containing 0.10 mole of Ba^{++}, only 0.0001 mole remains unprecipitated, so that essentially 0.10 mole of $Ba(IO_3)_2$ has precipitated. The precipitation reaction

$$Ba^{++} + 2IO_3^- \rightleftarrows Ba(IO_3)_2(\text{solid})$$

shows that two moles of IO_3^- are required per mole of Ba^{++} precipitated; 0.20 mole of KIO_3 will be required. An additional 0.0039 moles of IO_3^- must be added to keep the $[IO_3^-]$ at the level calculated in (4.12), and the total moles of KIO_3 needed is

$$0.200 + 0.0039 = 0.204 \text{ moles per liter}$$

4–3. FRACTIONAL PRECIPITATION

Problems related to the type just discussed are those involving fractional precipitation, that is, precipitation of certain specific ions from a solution. This technique is used for removing interferences in analytical methods and forms the basis of several qualitative analysis schemes.

Example 4.6. To a solution containing 0.010 M Ba^{++} and 0.010 M Ca^{++}, SO_4^{--} is added (as Na_2SO_4) in small increments.

(a) At what $[SO_4^{--}]$ will $BaSO_4$ start to precipitate?

As in Example 4.4

$$[Ba^{++}][SO_4^{--}] = 1.0 \times 10^{-10}$$

$$[SO_4^{--}] = \frac{1.0 \times 10^{-10}}{0.010} = 1.0 \times 10^{-8} \, M$$

for $BaSO_4$ to start to precipitate.

(b) At what $[SO_4^{--}]$ will $CaSO_4$ start to precipitate?

$$[Ca^{++}][SO_4^{--}] = 1.0 \times 10^{-5}$$

$$[SO_4^{--}] = \frac{1.0 \times 10^{-5}}{0.010} = 1.0 \times 10^{-3} \, M$$

for $CaSO_4$ to start to precipitate.

(c) What is the $[Ba^{++}]$ in solution when $CaSO_4$ starts to precipitate?
Using the $[SO_4^{--}]$ from (c) and the K_{sp} expression of $BaSO_4$

$$[Ba^{++}] = \frac{1.0 \times 10^{-10}}{1.0 \times 10^{-3}} = 1.0 \times 10^{-7} \, M$$

when $CaSO_4$ starts to precipitate.

(d) Over what $[SO_4^{--}]$ range can Ba^{++} be separated quantitatively from Ca^{++}?

To separate Ba^{++} quantitatively from Ca^{++}, less than 0.1 percent
of Ba^{++} must remain in solution, and no Ca^{++} must precipitate. Under
these conditions 99.9 percent of Ba^{++} is precipitated as $BaSO_4$ while all
of the Ca^{++} remains in solution and filtration at this stage will lead to
a successful separation. Since the $[Ba^{++}]$ must be less than

$$10^{-3} \times 0.010 \, M = 1.0 \times 10^{-5} \, M$$

$$[SO_4^{--}] \geqq \frac{1.0 \times 10^{-10}}{1.0 \times 10^{-5}} = 1.0 \times 10^{-5} \, M$$

to precipitate at least 99.9 percent of the Ba^{++}. But from (c) above,
the $[SO_4^{--}]$ must be less than $1.0 \times 10^{-3} \, M$ to prevent precipitation of
$CaSO_4$. Therefore the range of $[SO_4^{--}]$ for successful separation is

$$1.0 \times 10^{-5} \, M < [SO_4^{--}] < 1.0 \times 10^{-3} \, M$$

Concentrations of Ba^{++}, Ca^{++}, and SO_4^{--} at various stages of the
above problem are shown in Figure 4.2.

Example 4.7. A solution initially contains 0.010 M Pb^{++} and 0.010 M
Mn^{++}.

(a) At what $[S^{--}]$ will the $[Pb^{++}]$ be decreased to $1.0 \times 10^{-5} \, M$,
because of precipitation of PbS?

$$PbS(\text{solid}) \rightleftarrows Pb^{++} + S^{--}$$

$$[Pb^{++}][S^{--}] = 7 \times 10^{-29}$$

To decrease the $[Pb^{++}]$ to $1.0 \times 10^{-5} \, M$, the $[S^{--}]$ must be given by

$$[S^{--}] = \frac{7 \times 10^{-29}}{1.0 \times 10^{-5}} = 7 \times 10^{-24} \, M$$

(b) Will MnS precipitate from this solution under these conditions?

$$MnS(\text{solid}) \rightleftarrows Mn^{++} + S^{--}$$

$$[Mn^{++}][S^{--}] \rightleftarrows 7 \times 10^{-16}$$

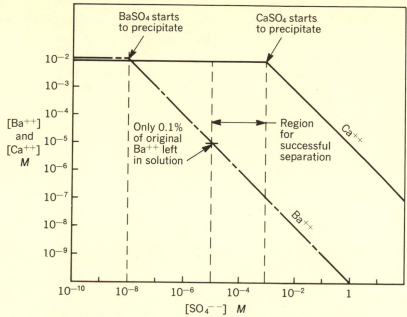

FIGURE 4.2. Fractional precipitation of $BaSO_4$ from a solution initially containing $0.010\ M\ Ba^{++}$ and $0.010\ M\ Ca^{++}$.

Comparing the ion concentration product with the K_{sp},

$$(1.0 \times 10^{-2})(7 \times 10^{-24}) \ll 7 \times 10^{-16}$$

we find that MnS will not precipitate.

If the $[S^{--}]$ in this solution is maintained at $7 \times 10^{-24}\ M$, Pb^{++} can be separated from Mn^{++} by filtration; the PbS remains on the filter paper, while the Mn^{++}, in solution, passes through. A technique for adjusting and maintaining the $[S^{--}]$, by control of the hydrogen ion concentration, will be described in section 4.5.

Separations made by fractional precipitation techniques will often not be as complete as the results of these calculations imply because of *coprecipitation*. Coprecipitation involves the carrying down of normally soluble substances by inclusion in the precipitate or by adsorption on the surface of the precipitate.

4–4. COMPETING ACID-BASE EQUILIBRIA

Frequently the ions which make up a precipitate can participate in other equilibria. For example, precipitates containing anion bases, such as CN^-, CrO_4^{--}, OH^-, S^{--}, etc., are more soluble in acidic solutions because of the competing reaction of the anions with H^+. In fact, because of these reactions, precipitates containing these anions will often be more soluble in water than a simple Type 1 calculation will predict.

Example 4.8. Calculate the $[Ag^+]$ and $[CrO_4^{--}]$, and the solubility of Ag_2CrO_4, in a solution containing an excess of solid Ag_2CrO_4 in which the final $[H^+]$ is 0.010 M.

In addition to the K_{sp} expression of Ag_2CrO_4

$$[Ag^+]^2[CrO_4^{--}] = 1.9 \times 10^{-12} \tag{4.13}$$

we must consider the reaction of CrO_4^{--} with H^+,

$$HCrO_4^- \rightleftarrows H^+ + CrO_4^{--}$$

$$\frac{[H^+][CrO_4^{--}]}{[HCrO_4^-]} = 3.2 \times 10^{-7} \tag{4.14}$$

The formation of H_2CrO_4 is unimportant because $HCrO_4^-$ is a negligibly weak base. Since all of the Ag^+ and all of the chromate species come from the Ag_2CrO_4 in this solution, and two Ag^+'s are liberated per one chromate, the material balance equation becomes

$$[Ag^+] = 2\{[CrO_4^{--}] + [HCrO_4^-]\} \tag{4.15}$$

Finally, with the equation

$$[H^+] = 0.010 \ M \tag{4.16}$$

we have supplied four equations for the four unknowns. From (4.14) and (4.16), we obtain,

$$[HCrO_4^-] = \frac{[H^+][CrO_4^{--}]}{3.2 \times 10^{-7}} = \frac{(0.010)\ [CrO_4^{--}]}{3.2 \times 10^{-7}} \tag{4.17}$$

Combining (4.15) and (4.17)

$$[Ag^+] = 2[CrO_4^{--}]\{1 + (3.1 \times 10^4)\} \tag{4.18}$$

$$[CrO_4^{--}] = 1.6 \times 10^{-5}\ [Ag^+] \tag{4.19}$$

Combining (4.13) and (4.19)

$$[Ag^+]^3 = \frac{1.9 \times 10^{-12}}{1.6 \times 10^{-5}} = 120 \times 10^{-9}$$

$$[Ag^+] = 4.9 \times 10^{-3}\ M \quad Answer$$

$$[CrO_4^{--}] = 7.8 \times 10^{-8}\ M \quad Answer$$

$$[HCrO_4^-] = 2.4 \times 10^{-3}\ M$$

Since two moles of Ag^+ are produced for every mole of Ag_2CrO_4 that dissolves, the solubility is equal to one-half of the $[Ag^+]$ in this case, or about 2.4×10^{-3} moles/liter. The increase in solubility of Ag_2CrO_4 over that calculated in Example 4.1 (for Ag_2CrO_4 in pure water) is caused by the reaction of CrO_4^{--} with H^+, which, by Le Chatelier's principle, shifts the solubility reaction, (4.8), to the right.

Example 4.9. What $[H^+]$ is needed to dissolve completely 0.0010 mole of AgCN in 1 liter of solution?

(1)
$$AgCN + H^+ \rightleftarrows Ag^+ + HCN \qquad (4.20)$$

(2)
$$[Ag^+][CN^-] = 1.6 \times 10^{-14} \qquad (4.21)$$

$$\frac{[H^+][CN^-]}{[HCN]} = 7.2 \times 10^{-10} \qquad (4.22)$$

(3)   Material balances: since 0.0010 mole of AgCN dissolves,

$$[Ag^+] = 0.0010 \; M \qquad (4.23)$$

$$[CN^-] + [HCN] = 0.0010 \; M \qquad (4.24)$$

We have now supplied four equations for four unknowns. From (4.21) and (4.23)

$$[CN^-] = 1.6 \times 10^{-11} \qquad (4.25)$$

so that from (4.24)

$$[HCN] = 1.0 \times 10^{-3} \qquad (4.26)$$

Finally, from (4.22) and the above results

$$[H^+] = \frac{(7.2 \times 10^{-10})(1.0 \times 10^{-3})}{(1.6 \times 10^{-11})} = 4.5 \times 10^{-2} \; M$$

Therefore the final $[H^+]$ needed in the solution to effect the solution of AgCN is 0.045 M. Note that the number of moles of H^+ which must be *added* is

$$[H^+] + [HCN] = 0.045 + 0.001 = 0.046 \text{ moles/liter}$$

Admittedly, problems involving both solubility and acid-base equilibria can be much more complicated than the preceding examples. When problems effectively state the concentrations of some of the species at equilibrium, such as H^+ in Example 4.8 or Ag^+ in Example 4.9, the solutions can usually be obtained quite easily. However, if none of the concentrations at equilibrium are known, mathematical difficulties may be much greater. For example the calculation of the solubility of

AgCN in water, without neglecting the basicity of the CN^-, although similar to Example 4.9, is much more difficult. Although the equations that are needed to solve the problem can be written easily, as the reader should demonstrate, approximations are not readily made, and the exact solution involves solving a fourth degree equation.

Calculations of the solubility of precipitates can also be complicated by complex ion formation of the metal ion of the precipitate. For example silver salts are more soluble in solutions containing NH_3 because of the formation of the complex ion $Ag(NH_3)_2^+$. Similarly, calculations of the solubility of AgCN in Example 4.9 neglected the formation of cyanosilver complexes, such as $Ag(CN)_2^-$. Consideration of complex ion equilibria, and some further refinements of equilibrium theory, will be discussed in Chapter 5.

4–5. *HYDROGEN SULFIDE SEPARATIONS*

A frequent application of solubility and acid-base equilibria is the separation of metal ions by precipitation of their sulfides at controlled pH's. Example 4.7 illustrated the principle of fractional precipitation of metals by controlling the $[S^{--}]$, and Pb^{++} was shown to be separable from Mn^{++} by maintaining the $[S^{--}]$ at $7 \times 10^{-24} M$. But how can one adjust and maintain a sulfide ion concentration at a certain level? In this case it is not too difficult, because sulfide ion is a weak base (or is an anion of a weak acid) and is involved in the equilibrium

$$H_2S \rightleftarrows 2H^+ + S^{--} \tag{4.27}$$

Furthermore, H_2S is only moderately soluble and can be maintained at a fixed concentration by bubbling H_2S gas continuously through a solution and forming a saturated solution of H_2S. The overall equilibrium constant expression for (4.27), calculated by multiplying the expressions for the stepwise ionization of H_2S, is*

---

* Although there is nothing wrong with forming an equilibrium constant expression by multiplying the expressions for two separate equilibria, the resulting "lumped" expression must be used with caution. (4.28) can be employed, as long as we know two of the three concentrations involved in it *at equilibrium* and want to calculate the third concentration. It would be unwise to use this expression in a Type 1 H_2S ionization problem however, because we might tend to neglect HS^-, which is present in the solution in significant amounts, and write an erroneous equation, such as

$$[H^+] = 2[S^{--}]$$

instead of the correct one (see Example 3.7).

$$\frac{[H^+]^2 [S^{--}]}{[H_2S]} = K_1K_2 = 1.1 \times 10^{-21} \qquad (4.28)$$

Since a saturated solution of H_2S at $25°$ C contains an $[H_2S]$ of about
0.1 M, we can write

$$[H^+]_{s.s.}^2.[S^{--}]_{s.s.} = 1.1 \times 10^{-22} \qquad (4.29)$$

(where the subscripts, s.s., remind us that this expression is only valid
for a solution saturated with H_2S). (4.29) indicates that the $[S^{--}]$ in a
solution is determined by the $[H^+]$, and that the $[S^{--}]$ can be adjusted
by variation of the pH. For example, to maintain an $[S^{--}]$ of 7×10^{-24}
M, as required in Example 4.7, the $[H^+]$ concentration, in a saturated
solution of H_2S, can be calculated from (4.29)

$$[H^+]_{s.s.} = \sqrt{\frac{1.1 \times 10^{-22}}{7 \times 10^{-24}}} = 4.0 \ M$$

Example 4.10. A solution contains 0.0010 M each of Cd^{++} and Zn^{++}.
We want to precipitate the Cd^{++} quantitatively as CdS (that is, decrease
the $[Cd^{++}]$ to about $1.0 \times 10^{-6} M$) and not precipitate any Zn^{++} as
ZnS. If the solution is maintained saturated with H_2S, over what range
of pH's can this separation be accomplished?

$$[Cd^{++}] [S^{--}] = 1.0 \times 10^{-28}$$

To decrease the $[Cd^{++}]$ to $1.0 \times 10^{-6} M$, the $[S^{--}]$ must be given by

$$[S^{--}] = \frac{1.0 \times 10^{-28}}{1.0 \times 10^{-6}} = 1.0 \times 10^{-22} \ M$$

$$[Zn^{++}] [S^{--}] = 1.6 \times 10^{-23}$$

To prevent precipitation of Zn^{++} from this solution, the $[S^{--}]$ must be
maintained below the value calculated as

$$[S^{--}] = \frac{1.6 \times 10^{-23}}{1.0 \times 10^{-3}} = 1.6 \times 10^{-20} \ M$$

Therefore, we wish to maintain the $[S^{--}]$ in the following range

$$1.0 \times 10^{-22} \ M < [S^{--}] < 1.6 \times 10^{-20} \ M$$

From (4.29), to maintain the $[S^{--}]$ greater than $1.0 \times 10^{-22} M$ in a
solution saturated with H_2S, we must maintain the $[H^+]$ less than the
amount calculated as

$$[H^+]_{s.s.} = \sqrt{\frac{1.1 \times 10^{-22}}{1.0 \times 10^{-22}}} = 1.1 \ M$$

To maintain the $[S^{--}]$ less than 1.6×10^{-20} M, we must maintain the $[H^+]$ greater than the amount calculated as

$$[H^+]_{s.s.} = \sqrt{\frac{1.1 \times 10^{-22}}{1.6 \times 10^{-20}}} = 0.083 \ M$$

Therefore, for a successful separation, the $[H^+]$ must be maintained in the range

$$1.1 \ M > [H^+]_{s.s.} > 0.083$$

or

$$0 < pH < 1.1 \quad \textit{Answer}$$

PROBLEMS

4.1. Calculate the K_{sp} for each of the following from the given data.
  (a) MX_2, which has a solubility of 0.0020 moles/liter.
  (b) M_2X_3, molecular weight 150, which has a solubility of 0.045 mg per 100 ml.
  (c) M_2X, which produces a solution with an X^{--} concentration of 2.0×10^{-5} M when an excess of M_2X is equilibrated with water.
  (d) MX_3, which yields a solution with an M^{+++} concentration of 2.0×10^{-10} M when excess solid MX_3 is equilibrated with an aqueous solution of 0.10 M NaX.

4.2. From the tabulated K_{sp}'s calculate the solubility of the following substances in moles/liter and mg per 100 ml. Calculate the concentrations of all of the ions present in a solution prepared by equilibrating an excess of the solid with water.
  (a) $BaSO_4$ (b) Ag_2SO_4 (c) CaF_2
  (d) Hg_2Cl_2 (e) $Ce(IO_3)_2$ (f) $Cu(IO_3)_2$
  (g) $AgIO_3$ (h) $Mg(OH)_2$ (i) $PbBr_2$

4.3. Perform the same calculations as those in 4.2 on the following (assume hydrolysis of the anion does not occur to a significant extent).
  (a) $Ba_3(AsO_4)_2$ (b) FeS   (c) $Pb(OH)_2$
  (d) Ag_3PO_4 (e) Tl_2S (f) $Al(OH)_3$

4.4. Calculate, for the following solutions, the concentration of the anion necessary to (1) just start precipitation of the indicated metal ion and (2) decrease the concentration of the metal ion to 0.1 percent of its initial concentration.

(a) $0.010\ M$ Ba^{++}, precipitated as $BaSO_4$

(b) $0.10\ M$ Mg^{++}, precipitated as MgF_2

(c) $0.010\ M$ Hg_2^{++}, precipitated as $Hg_2(CNS)_2$

(d) $0.10\ M$ Mg^{++}, precipitated as $Mg(OH)_2$

(e) $0.010\ M$ Ca^{++}, precipitated as $Ca_3(AsO_4)_2$

4.5. Calculate the concentration of the ions in the following solutions and the solubilities of the indicated precipitates in moles/liter.

(a) $AgBr$, in a solution of $0.10\ M$ $NaBr$

(b) $AgBr$, in a solution of $0.50\ M$ $AgNO_3$

(c) BaF_2, in a solution of $0.010\ M$ NaF

(d) BaF_2, in a solution of $0.10\ M$ $Ba(NO_3)_2$

(e) $Ce(IO_3)_3$, in a solution of $0.050\ M$ $NaIO_3$

(f) $Ce(IO_3)_3$, in a solution of $0.050\ M$ $Ce(NO_3)_3$

(g) $Ag(CH_3COO)$, in a solution of $0.10\ M$ CH_3COONa

4.6. Calculate the concentrations of the ions in the solutions that result when the following solutions are mixed, precipitation has occurred, and equilibrium attained.

(a) 200 ml $0.10\ M$ KI and 300 ml $0.050\ M$ $AgNO_3$

(b) 100 ml $0.20\ M$ $CaCl_2$ and 100 ml $0.050\ M$ NaF

(c) 400 ml of $0.050\ M$ $Ce(NO_3)_3$ and 600 ml $0.30\ M$ KIO_3

(d) 100 ml $1.0\ M$ $AgNO_3$ and 300 ml $1.0\ M$ K_2SO_4

(e) 100 ml $0.10\ M$ $Ba(OH)_2$ and 100 ml $0.50\ M$ $MgSO_4$ (Note: two precipitates form.)

4.7. A solution contains $0.010\ M$ Ba^{++} and $0.010\ M$ Ca^{++}. Solid NaF is added in very, very small increments. (a) Calculate the F^- concentration necessary to start precipitation of BaF_2. (b) Calculate the F^- concentration necessary to start precipitation of CaF_2. (c) Which precipitates first, BaF_2 or CaF_2? (d) What is the concentration of the metal ion that precipitates first when the second ion just starts to precipitate?

4.8. Make calculations such as those in Problem 4.7, for a solution of $0.010\ M$ Tl^+ and $0.010\ M$ Pb^{++}, to which $NaIO_3$ is added.

4.9. Calculate for the following (1) the hydrogen ion concentration in the final solution and (2) the number of moles of H^+ (as, for example, HNO_3) that must be added to dissolve completely, in 1 liter of solution the following precipitates:

(a) 0.010 mole $AgOH$ (b) 0.050 mole SrF_2

(c) 0.010 mole $Fe(OH)_3$ (d) 0.0010 mole $Hg_2(CN)_2$

4.10. Calculate the solubility (in moles/liter) of the substances in Problem 4.3 in solutions whose final (equilibrium) pH is (1) 2.00 (2) 4.00 (3) 7.00.

4.11. How many grams of $Fe(OH)_3$ will dissolve in 2.0 liters of (a) water (b) 0.010 M $Fe(NO_3)_3$ (c) 0.10 M NaOH (d) 0.001 M HCl?

4.12. The separation of Al^{+++} and Mg^{++} in limestone analysis is accomplished by precipitation of $Al(OH)_3$, leaving Mg^{++} in solution. How many grams of NH_4Cl must be added to 100 ml of a solution 0.010 M Al^{+++} and 0.010 M Mg^{++}, and 0.10 M NH_3 to just prevent the precipitation of $Mg(OH)_2$?

4.13. What pH range, if any, will be useful for the quantitative separation of the following metals in a solution saturated (0.10 M) in H_2S?

  (a) 0.010 M Sn^{++} and 0.010 M Mn^{++}
  (b) 0.010 M Zn^{++} and 0.10 M Ni^{++}
  (c) 0.10 M Pb^{++} and 0.10 M Fe^{++}
  (d) 0.010 M Cu^{++} and 0.010 M Co^{++}

SUPPLEMENTARY READING

Blake, R. F., Demonstration of Dynamic Equilibrium Using Radioactive Iodine, *J. Chem. Educ.*, **33**, 354 (1956).

Butler, J. N., Calculation of Molar Solubilities From Equilibrium Constants, *J. Chem. Educ.*, **38**, 460 (1961).

Petrucci, R. H., and P. C. Moews, The Precipitation and Solubility of Metal Sulfides, *J. Chem. Educ.*, **39**, 391 (1962).

CHAPTER 5 COMPLEX ION EQUILIBRIA

5-1. COMPLEX IONS

METAL IONS AND LIGANDS

A metal ion in solution is usually associated with one or more groups that stabilize the ion and keep it in solution. This species, composed of the metal ion and its associated groups (*ligands*), is called a *complex ion*. The metals that form complex ions are usually thought of as metallic, such as Fe^{+++}, Cd^{++}, Zn^{++}, Cu^{++}, and Pt^{++++}. The alkali metals, Na^+, K^+, and Li^+, form practically no complexes, while the alkaline earth metals, Ca^{++}, Ba^{++}, and Sr^{++}, form relatively few complex ions. The groups that act as ligands (sometimes called *complexing agents*) are usually neutral or anionic species which have a free pair of electrons that may be "donated" to the metal ion. Some familiar ligands include NH_3, Cl^-, F^-, and CN^-. The formation of typical complex ions is illustrated by the following reactions.

$$Cu^{++} + 4NH_3 \rightleftarrows Cu(NH_3)_4^{++} \tag{5.1}$$

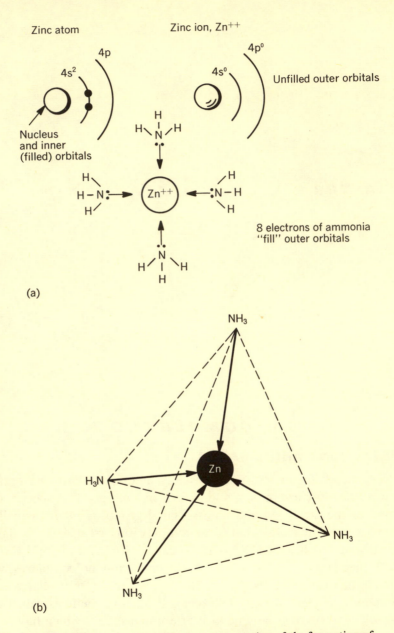

FIGURE 5.1. (a) Schematic representation of the formation of the $Zn(NH_3)_4^{++}$ complex ion. (b) Diagram suggesting geometrical arrangement of ammonia ligands about the central zinc ion. The ammonias form a tetrahedron about the zinc. Complexes containing six ligands usually have an octahedral structure.

$$Ag^+ + 2CN^- \rightleftharpoons Ag(CN)_2^- \qquad (5.2)$$

$$Fe^{+++} + 6F^- \rightleftharpoons FeF_6^{---} \qquad (5.3)$$

A schematic representation of the formation of a complex ion is shown in Fig. 5.1.

Water itself will act as a ligand and the metal ions will often be in the form of *aquo-complexes*. Those species usually written as uncomplexed metal ions in chemical equations, such as Cu^{++}, are really associated with water molecules, and should more exactly be written as the aquo-complexes, i.e., $Cu(H_2O)_4^{++}$. We will continue to write these aquo-complexes as the bare metal ion (just as we represent the aquated proton as H^+) remembering that water molecules are really associated with the metal ion. Complexation reactions, such as (5.1), actually represent the conversion of the aquo-complex to a more stable complex ion, i.e.,

$$Cu(H_2O)_4^{++} + 4NH_3 \rightleftharpoons Cu(NH_3)_4^{++} + 4H_2O \qquad (5.4)$$

MAXIMUM COORDINATION NUMBER

How can one predict the number of ligands associated with a metal ion? Actually there are no hard and fast rules to allow us to predict this. A convenient "rule-of-thumb" is that the maximum number of ligands that can associate with a metal ion (the so-called *maximum coordination number*) is very often twice the ionic charge on the metal ion. Therefore the maximum coordination number of Ag^+ is 2, of Cu^{++}, Cd^{++}, or Zn^{++} is 4, and of Fe^{+++}, Al^{+++}, or Co^{+++} is 6. There are many exceptions to this guide. Ions which commonly show two oxidation states will usually assume the coordination number of the higher oxidation state, so that Fe^{++} (which also exists as Fe^{+++}) and Co^{++} (which also exists as Co^{+++}) frequently show maximum coordination numbers of 6. Furthermore, maximum coordination numbers larger than 6 are very rare, and metals which are quadrivalent, such as Pt^{++++} and Sn^{++++} generally have maximum coordination numbers of 6. The charge exhibited by the complex ion is determined by balancing the ionic equation for its formation. The only unequivocal way of finding out the composition of a complex ion (assuming it has been previously studied) is to consult a table of complex ions, such as the one in Appendix C, and find out what has been discovered experimentally.

Actually a metal may not necessarily be associated with its maximum number of ligands (other than water). For example a solution containing

Cu^{++} and NH_3 contains all of the following species: $Cu(NH_3)_4^{++}$, $Cu(NH_3)_3^{++}$, $Cu(NH_3)_2^{++}$, $Cu(NH_3)^{++}$, and Cu^{++}.* The relative concentrations of these species depend upon the concentrations of Cu^{++} and NH_3 in the solution. Since equilibrium constant expressions can be written for the formation of each of these species, and frequently many of these equilibrium constants are quite near one another in magnitude, the solution of complex ion problems may be quite difficult. On the other hand, if there is a moderate excess of complexing agent in the solution, the most highly coordinated species will predominate, and lower complexes can often be neglected. For example, in a solution prepared by adding 0.10 mole of Cu^{++} and 2.0 moles of NH_3 to a liter of water, almost all the copper will be in the form of $Cu(NH_3)_4^{++}$. In the problems that follow, we will deal with solutions containing an excess of complexing agent. We will, therefore, only consider the most highly coordinated species in equilibrium with the bare metal ion and hence only discuss common ion effect (Type 2) problems. Problems involving the intermediate complexes in solutions containing varying concentrations of complexing agents are considered in Chapter 8.

5–2. COMPLEX ION PROBLEMS

The formation of a complex ion from its ions, such as

$$Cd^{++} + 4CN^- \rightleftarrows Cd(CN)_4^{--} \tag{5.5}$$

is a reaction that attains equilibrium. An equilibrium constant expression, such as

$$\frac{[Cd(CN)_4^{--}]}{[Cd^{++}][CN^-]^4} = K_{stab} = 7.1 \times 10^{18} \tag{5.6}$$

can be written. K_{stab} is the *stability constant* of the complex ion (which is sometimes written K_f, the formation constant).** A large K_{stab} indicates a complete formation (or a small dissociation) of the complex ion.

* Actually $Cu(NH_3)_4^{++}$, $Cu(NH_3)_3(H_2O)^{++}$, $Cu(NH_3)_2(H_2O)_2^{++}$, $Cu(NH_3)(H_2O)_3^{++}$, and $Cu(H_2O)_4^{++}$. At very high concentrations of free NH_3 some $Cu(NH_3)_5^{++}$ also exists.

** (5.5) is sometimes written as the dissociation of the complex ion

$$Cd(CN)_4^{--} \rightleftarrows Cd^{++} + 4CN^-$$

and the equilibrium constant expression written in terms of the *instability constant* or *dissociation constant*, K_d, where

$$K_d = 1/K_{stab}$$

We will use the conventions and expressions shown in (5.5) and (5.6) in this book.

Example 5.1. Calculate the concentrations of the various species at equilibrium in a solution prepared by diluting 2.40 moles of KCN and 0.10 mole of $Cd(NO_3)_2$ to 1 liter with water.

(1) $$Cd^{++} + 4CN^- \rightleftarrows Cd(CN)_4^{--}$$

The salts $Cd(NO_3)_2$ and KCN are completely ionized.

(2) $$\frac{[Cd(CN)_4^{--}]}{[Cd^{++}][CN^-]^4} = 7.1 \times 10^{18}$$

(3) Material balance equations
Material balance for cadmium:

$$[Cd^{++}] + [Cd(CN)_4^{--}] = 0.10 \tag{5.7}$$

Material balance for cyanide:

$$[CN^-] + 4[Cd(CN)_4^{--}] = 2.40 \tag{5.8}$$

The 4 is put before the $[Cd(CN)_4^{--}]$ term because each mole of $Cd(CN)_4^{--}$ contains 4 moles of CN^-. (5.6), (5.7), and (5.8) represent 3 equations which can be solved for the 3 unknowns: $[Cd^{++}]$, $[CN^-]$, and $[Cd(CN)_4^{--}]$.* Since K_{stab} is so large, most of the cadmium is in the form of the complex ion, so the approximation

$$[Cd(CN)_4^{--}] \gg [Cd^{++}]$$

can be made. Therefore

$$[Cd(CN)_4^{--}] \approx 0.10 \ M$$

$$[CN^-] \approx 2.40 - 4(0.10) \approx 2.00 \ M$$

Using these expressions and the K_{stab} expression, we obtain

$$\frac{(0.10)}{[Cd^{++}](2.00)^4} = 7.1 \times 10^{18}$$

$$[Cd^{++}] = 8.8 \times 10^{-22} \ M \quad \textit{Answer}$$

---

* Alternately, the electroneutrality equation:

$$[K^+] + 2[Cd^{++}] = [CN^-] + [NO_3^-] + 2[Cd(CN)_4^{--}]$$

and the material balance equations:

$$[K^+] = 2.40 \ M \quad [NO_3^-] = 0.20 \ M$$

can be written in place of (5.7) or (5.8). Note that combination of these equations with (5.8) for example leads to (5.7).

5-3. COMPLEX IONS AND COMPETING EQUILIBRIA

Complex ion formation offers a method for controlling the concentration of a metal ion in solution. We can prevent a metal ion from precipitating or taking part in a reaction many times by forming a complex ion and decreasing the concentration of metal ion in the solution.

PRECIPITATION SEPARATIONS

Example 5.2. A solution contains $0.010\ M$ each of Ni^{++} and Zn^{++}. If KCN is added to the solution until the $[CN^-]$ is $1.0\ M$, and the $[S^{--}]$ is maintained at $0.50\ M$, what amount of the Ni^{++} and Zn^{++} will remain unprecipitated?

(a) Let us first assume that no NiS or ZnS precipitates.

(1) Addition of CN^- causes the formation of the cyano-complexes of Ni^{++} and Zn^{++}

$$Ni^{++} + 4CN^- \rightleftarrows Ni(CN)_4{}^{--} \tag{5.9}$$

$$Zn^{++} + 4CN^- \rightleftarrows Zn(CN)_4{}^{--} \tag{5.10}$$

(2)
$$\frac{[Ni(CN)_4{}^{--}]}{[Ni^{++}][CN^-]^4} = 1.0 \times 10^{22} \tag{5.11}$$

$$\frac{[Zn(CN)_4{}^{--}]}{[Zn^{++}][CN^-]^4} = 8.3 \times 10^{17} \tag{5.12}$$

(3)
$$[CN^-] = 1.0\ M \tag{5.13}$$

$$[Ni(CN)_4{}^{--}] + [Ni^{++}] = 0.010\ M \tag{5.14}$$

$$[Zn(CN)_4{}^{--}] + [Zn^{++}] = 0.010\ M \tag{5.15}$$

Since the stability constants of both complexes are so large, most of the metals will be in the form of their complexes, that is

$$[Ni(CN)_4{}^{--}] \approx 0.010\ M \tag{5.16}$$

$$[Zn(CN)_4{}^{--}] \approx 0.010\ M \tag{5.17}$$

The free metal ion concentrations can now be calculated from (5.11), (5.13) and (5.16), and (5.12), (5.13), and (5.17).

$$[Ni^{++}] = \frac{(0.010)}{(1.0)^4 \times (1.0 \times 10^{22})} = 1.0 \times 10^{-24}\ M$$

$$[Zn^{++}] = \frac{(0.010)}{(1.0)^4 \times (8.3 \times 10^{17})} = 1.2 \times 10^{-20} \, M$$

(b) Now let us see if NiS or ZnS precipitates.
From K_{sp} expressions

$$[Ni^{++}][S^{--}] = 1.0 \times 10^{-24} \qquad (5.18)$$

$$[Zn^{++}][S^{--}] = 1.6 \times 10^{-23} \qquad (5.19)$$

and

$$[S^{--}] = 0.50 \, M \qquad (5.20)$$

we observe two things: NiS does not precipitate, since the ion concentration product is less than the K_{sp} of NiS

$$(1.0 \times 10^{-24})(0.50) < 1.0 \times 10^{-24}$$

so that 100 percent of the Ni^{++} remains unprecipitated; and ZnS precipitates, since the ion concentration product here is greater than the K_{sp}

$$(1.2 \times 10^{-20})(0.50) > 1.6 \times 10^{-23}$$

The amount of zinc left unprecipitated can now be calculated. At the existing $[S^{--}]$ from (5.19),

$$[Zn^{++}] = \frac{1.6 \times 10^{-23}}{0.50} = 3.2 \times 10^{-23} \, M$$

Zinc is also present in the form of the complex ion, $Zn(CN)_4^{--}$. From (5.12)

$$[Zn(CN)_4^{--}] = 8.3 \times 10^{17} \times 3.2 \times 10^{-23} \times (1.0)^4$$

$$[Zn(CN)_4^{--}] = 2.7 \times 10^{-5} \, M$$

Therefore the percentage of zinc left unprecipitated is given by

$$\%Zn \text{ unppted.} = \frac{(2.7 \times 10^{-5}) + (3.2 \times 10^{-23})}{1.0 \times 10^{-2}} \times 100 = 0.27\% \quad Answer$$

Under these conditions a good separation of Zn and Ni could be made.

DISSOLUTION OF PRECIPITATES

In Chapter 4 we noted that precipitates containing an anion base could be dissolved in acidic solutions. Similarly, many precipitates containing metal ions can be dissolved by formation of a complex ion. *Example 5.3.* How many moles of NH_3 must be added per liter to dissolve completely 0.010 mole of AgBr?

(1) $$AgBr + 2NH_3 \rightleftarrows Ag(NH_3)_2{}^+ + Br^- \quad (5.21)$$

(2) $$[Ag^+][Br^-] = 5.0 \times 10^{-13} \quad (5.22)$$

$$\frac{[Ag(NH_3)_2{}^+]}{[Ag^+][NH_3]^2} = 1.7 \times 10^7 \quad (5.23)$$

(3) Since 0.010 mole of AgBr dissolves, and all of the silver and bromide in the solution come from this source, the total bromide concentration, and the total silver concentration in the solution must be 0.010 M, that is,

$$[Br^-] = 0.010 \; M \quad (5.24)$$

$$[Ag^+] + [Ag(NH_3)_2{}^+] = 0.010 \; M \quad (5.25)$$

Since K_{stab} is large, $[Ag(NH_3)_2{}^+] \gg [Ag^+]$, so that

$$[Ag(NH_3)_2{}^+] \approx 0.010 \; M$$

(4) From the K_{sp} expression (5.22) and from (5.24)

$$[Ag^+] = \frac{5.0 \times 10^{-13}}{0.010} = 5.0 \times 10^{-11} \; M$$

Putting this value, and the value for $[Ag(NH_3)_2{}^+]$ into the K_{stab} expression (5.23)

$$[NH_3]^2 = \frac{0.010}{1.7 \times 10^7 \times 5.0 \times 10^{-11}} = 12$$

$$[NH_3] = 3.5 \; M$$

Therefore to dissolve 0.010 mole of AgBr, we must add $3.5 + 2(0.010) = 3.52$ moles of NH_3 per liter.

A problem similar to the one just solved, but which involves somewhat different considerations is one in which the concentration of complexing agent is given, and the solubility of a precipitate is to be calculated.

Example 5.4. Calculate the solubility of AgBr in a solution containing 1.0 mole of NH_3 per liter of solution.

The K_{sp} and K_{stab} expressions in this problem are (5.22) and (5.23). Since one silver ion is produced every time one bromide ion is produced, the concentration of all forms of silver in the solution must equal the bromide ion concentration

$$[Ag^+] + [Ag(NH_3)_2{}^+] = [Br^-] \quad (5.26)$$

Since 1.0 mole of NH_3 was added to the solution, all forms of NH_3 in solution must have a total concentration of 1.0 M

$$[NH_3] + 2[Ag(NH_3)_2{}^+] = 1.0 \ M \qquad (5.27)$$

(2 is written in front of the $[Ag(NH_3)_2{}^+]$ term because each mole of the complex contains two moles of NH_3).

The problem can be solved more easily if approximations are made in the above material balance equations. Which concentration is larger in (5.27) depends upon how soluble AgBr is in this medium. Guessing that the solubility is small,

$$[NH_3] \gg 2[Ag(NH_3)_2{}^+]$$

so that

$$[NH_3] \approx 1.0 \ M$$

From the K_{stab} expression (5.23)

$$[Ag(NH_3)_2{}^+] = 1.7 \times 10^7 \times (1.0)^2 \ [Ag^+]$$

Putting this value, and the value for $[Br^-]$ from the K_{sp} expression (5.22), into (5.26)

$$[Ag^+] + (1.7 \times 10^7) \ [Ag^+] = \frac{5.0 \times 10^{-13}}{[Ag^+]}$$

$$[Ag^+]^2 = \frac{5.0 \times 10^{-13}}{1.7 \times 10^7} = 2.9 \times 10^{-20}$$

$$[Ag^+] = 1.7 \times 10^{-10} \ M$$

$$[Ag(NH_3)_2{}^+] = 1.7 \times 10^7 \times 1.7 \times 10^{-10}$$

$$[Ag(NH_3)_2{}^+] = 2.9 \times 10^{-3} \ M$$

$$[Br^-] = \frac{5.0 \times 10^{-13}}{1.7 \times 10^{-10}} = 2.9 \times 10^{-3} \ M$$

The solubility of AgBr in a solution essentially 1.0 M in NH_3 is 2.9×10^{-3} moles per liter. *Answer*

Note that the approximation is valid, since

$$(1.0) \gg 5.8 \times 10^{-3}$$

Let us examine what would have happened if we had made a wrong guess during the approximation stage. Suppose we guessed that the solubility of AgBr is large in this solution, so that most of the NH_3

would be consumed in complex ion formation. The assumption then is

$$[Ag(NH_3)_2{}^+] \gg [NH_3]$$

so that

$$[Ag(NH_3)_2{}^+] \approx \tfrac{1}{2}(1.0) \approx 0.50 \; M$$

From (5.26), again using the K_{sp} expression (5.22),

$$[Ag^+] + 0.50 = \frac{5.0 \times 10^{-13}}{[Ag^+]}$$

$$[Ag^+] = 1.0 \times 10^{-12} \; M$$

Putting these values for $[Ag(NH_3)_2{}^+]$ and $[Ag^+]$ into the K_{stab} expression (5.23), and solving for $[NH_3]$

$$[NH_3]^2 = \frac{0.50}{1.7 \times 10^7 \times 1.0 \times 10^{-12}} = 2.9 \times 10^4$$

$$[NH_3] = 170 \; M$$

Obviously the approximation was not a good one, and either the opposite approximation should have been made (as at first), or if neither approximation works the problem must be solved rigorously, at the expense of some mathematical labor.

5–4. EFFECT OF pH ON LIGAND CONCENTRATION

Most ligands are bases, capable of reacting with hydrogen ion as well as with metal ions. In considering the formation of a metal complex ion, the extent of reaction of the ligand with H^+ must be taken into account. For example, consider the ligand CN^-. Cyano-complexes will be more extensively dissociated in acidic solutions because of the reaction of CN^- with H^+ (Le Chatelier's Principle). The extent of the reaction of CN^- with H^+ is described by the K_a expression

$$\frac{[H^+][CN^-]}{[HCN]} = K_a \tag{5.28}$$

Although we could solve problems of this type by the methods we have been using, let us formulate a slightly different approach frequently used in solving complex ion problems.

If the total number of moles of cyanide added to a solution is C_T, then

$$[CN^-] + [HCN] = C_T \qquad (5.29)$$

We can calculate the concentration of free CN^- at any pH by using (5.28) and (5.29)

$$C_T = [CN^-] + \frac{[H^+][CN^-]}{K_a} = [CN^-]\left[\frac{K_a + [H^+]}{K_a}\right]$$

Calling the fraction of the total cyanide in the form of CN^-, α_1, we have*

$$\alpha_1 = \frac{[CN^-]}{C_T} = \frac{K_a}{K_a + [H^+]} \qquad (5.30)$$

and

$$[CN^-] = \alpha_1 C_T \qquad (5.31)$$

Since, for a given acid, α_1 depends only upon the $[H^+]$, we can always calculate the $[CN^-]$ at a given pH by using (5.31). For example K_a for HCN is 7.2×10^{-10}, so that at a pH of 9.00

$$\alpha_1 = \frac{7.2 \times 10^{-10}}{1.0 \times 10^{-9} + 7.2 \times 10^{-10}} = 0.42$$

In a solution containing 0.10 mole of NaCN per liter at a pH of 9.00, the free $[CN^-]$ is

$$[CN^-] = 0.42(0.10) = 0.042 \ M$$

We can calculate α_1 at several pH's by this method and plot a curve showing α_1 for the cyanide system as a function of pH (Fig. 5.2).

Although the above calculations dealt with cyanide ion, relations such as these will obviously hold for any ligand which will react with only one hydrogen ion, and expressions such as (5.30) can be used for for NH_3, F^-, etc.

Example 5.5. Calculate the concentration of uncomplexed Cd^{++} in a solution prepared by diluting 2.40 moles of KCN and 0.10 mole of $Cd(NO_3)_2$ to 1 liter; the solution being adjusted to a pH of 9.00.

Except for the specifications of pH, this problem is very similar to Example 5.1. Proceeding as in that example we find

$$[Cd(CN)_4{}^{--}] = 0.10 \ M \quad C_T = 2.00 \ M^{**}$$

---

* The α notation is used to describe the fraction of an acid in a given form. For example, for the diprotic acid, H_2X, under given conditions, the fraction present as undissociated H_2X is called α_0, the fraction present as HX^- is called α_1, and the fraction present as X^{--} is called α_2.

** C_T here refers to all of the cyanide in the solution except that in the complex ion.

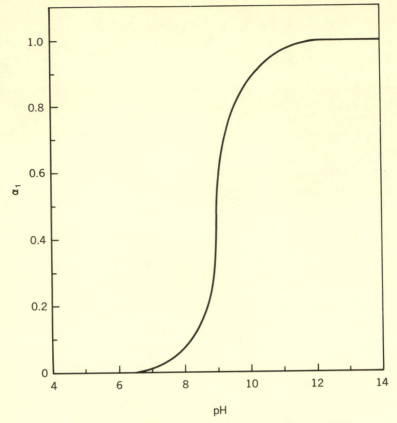

FIGURE 5.2. The fraction of total acid present as CN^-, α_1, a function of pH.

From (5.30) and (5.31), as previously, at a pH of 9.00, $\alpha_1 = 0.42$, so that

$$[CN^-] = 0.42(2.00) = 0.84 \ M$$

The $[Cd^{++}]$ is calculated from the stability constant expression (5.6)

$$[Cd^{++}] = \frac{[Cd(CN)_4^{--}]}{7.1 \times 10^{18}[CN^-]^4} = \frac{0.10}{7.1 \times 10^{18}(0.84)^4}$$

$$[Cd^{++}] = 2.8 \times 10^{-20} \ M \quad \textit{Answer}$$

Note that this free $[Cd^{++}]$ is somewhat higher than in Example 5.1 because of the more acidic pH. If the solution were made acidic enough, the complex would not even form, because α_1 would be essentially zero (Fig. 5.2) and the $[CN^-]$ would be extremely small.

5–5. CHELATES

POLYDENTATE LIGANDS

The ligands considered so far could satisfy only one coordination position of a metal ion per ligand, that is, each ligand had one pair of electrons that could be used in forming a metal-ligand bond. These ligands, such as ammonia ($:NH_3$), water ($H_2O:$), fluoride ion ($F:^-$) and cyanide ion ($:CN^-$), are called *monodentate* (literally " one-toothed ") ligands. Substances that contain several complexing groups on one molecule are called *polydentate* ligands. For example, when two ammonia molecules are tied back-to-back with a short carbon chain, ethylenediamine

$$
\begin{array}{cc}
H & H \\
:N—CH_2CH_2—N: \\
H & H
\end{array}
$$

is produced. This molecule contains two positions for forming bonds with a metal ion (one pair of electrons on each nitrogen), and forms a ring in the process of complexation of a metal ion. The ethylenediamine complex of Cu^{++} has the following structure:

$$
\begin{array}{ccc}
H & & H \\
N: & & :N \\
/H\backslash & & /H\backslash \\
CH_2 & & CH_2 \\
| & Cu^{++} & | \\
CH_2 & & CH_2 \\
\backslash H/ & & \backslash H/ \\
N: & & :N \\
H & & H
\end{array}
$$

Complexes formed by polydentate ligands are called *chelates* (" claw " compounds). The equilibria involved in chelate formation, and associated problems, are similar to other complex ion equilbria and problems already described. Thus for the copper-ethylenediamine complex (abbreviating ethylenediamine by en) we have

$$Cu^{++} + 2en \rightleftarrows Cu(en)_2^{++} \tag{5.32}$$

$$\frac{[Cu(en)_2^{++}]}{[Cu^{++}][en]^2} = 4 \times 10^{19} \tag{5.33}$$

Chelate complexes are usually more stable than those involving monodentate ligands. Generally, the greater the number of rings in a complex,
the more stable it is. Compare, for example, the stability constant of
$Cu(NH_3)_4^{++}$ (4×10^{12}) with that of $Cu(en)_2^{++}$ (4×10^{19}).

EDTA

Chelate compounds have been investigated quite vigorously during
the past twenty years, and a number of interesting polydentate ligands
have been synthesized. One of the most important of these is ethylene-
diaminetetraacetic acid, usually called EDTA,

Because it has so many coordinating sites available on the molecule
(on the nitrogen and oxygen atoms), EDTA always reacts in a simple
one-to-one ratio with a metal ion

$$M^{+n} + Y^{-4} \rightleftarrows MY^{n-4} \qquad (5.34)$$

(where Y^{-4} is an abbreviation for the ionized form of EDTA). For
example, the reaction of Cu^{++} with EDTA anion is given by the
equation

$$Cu^{++} + Y^{-4} \rightleftarrows CuY^{--} \qquad (5.35)$$

and yields the K_{stab} expression

$$\frac{[CuY^{--}]}{[Cu^{++}][Y^{-4}]} = 6.3 \times 10^{18} \qquad (5.36)$$

EDTA is such a good complexing agent that it coordinates the alkaline
earth metals, such as Ca^{++}, Ba^{++}, and Sr^{++}, and even forms weak
complexes with Li^+ and Na^+.

EDTA Problems

Problems involving polydentate ligands, such as EDTA, are actually simpler than those involving monodentate ones. Because there is only one possible complex species formed, no intermediate species need be considered, even in a rigorous treatment.

Example 5.6. A solution containing 0.010 M Ca^{++} is made 1.0 M in EDTA and 0.10 M in $C_2O_4^{--}$ (oxalate ion). Will CaC_2O_4 precipitate from this medium?

(a) First, let us assume that CaC_2O_4 does not precipitate.

(1) $$Ca^{++} + Y^{-4} \rightleftarrows CaY^{--} \tag{5.37}$$

(2) $$\frac{[CaY^{--}]}{[Ca^{++}][Y^{-4}]} = 5.0 \times 10^{10} \tag{5.38}$$

(3) $$[Ca^{++}] + [CaY^{--}] = 0.010 \ M \tag{5.39}$$

or since the complex is quite stable

$$[CaY^{--}] \approx 0.010 \ M \tag{5.40}$$

$$[Y^{-4}] = 1.0 \ M \tag{5.41}$$

Therefore, from (5.38), (5.40), and (5.41)

$$[Ca^{++}] = \frac{0.010}{5.0 \times 10^{10} \times 1.0} = 2.0 \times 10^{-13} \ M$$

(b) Now let us calculate the ion concentration product to see if CaC_2O_4 precipitates.

(1) $$CaC_2O_4 \rightleftarrows Ca^{++} + C_2O_4^{--} \tag{5.42}$$

(2) $$[Ca^{++}][C_2O_4^{--}] = 1.3 \times 10^{-9} \tag{5.43}$$

From the K_{sp} expression, (5.42), noting that

$$[C_2O_4^{--}] = 0.10 \ M$$

$$(2.0 \times 10^{-13})(0.10) < 1.3 \times 10^{-9}$$

No CaC_2O_4 precipitates. *Answer*

Actually, since Y^{-4} is the anion of EDTA, a tetraprotic acid, the actual concentration of Y^{-4} in a solution is dependent upon the pH. Using the values for the four ionization constants of EDTA (i.e., H_4Y), we can calculate, as in Section 5.4, a curve giving the fraction of the EDTA in the Y^{-4} form, (α_4), as a function of the pH (Fig. 5.3). In a problem in which the pH is specified, this curve can be used to find the actual concentration of Y^{-4} from the total concentration of EDTA,

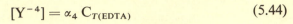

$$[Y^{-4}] = \alpha_4 \, C_{T(EDTA)} \qquad\qquad (5.44)$$

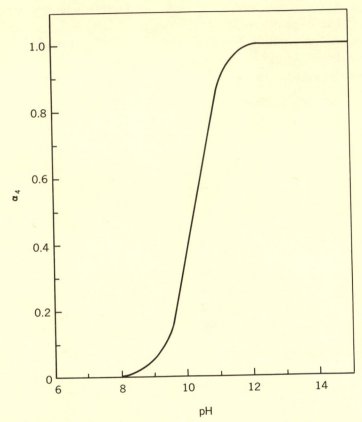

FIGURE 5.3. Fraction of the total EDTA (H_4Y) present in the form of the completely ionized species, Y^{-4}, α_4, as a function of pH.

EDTA has been used to effectively decrease the free metal ion concentration in solution by complexation in situations where the metal ions might interfere. EDTA is useful as a titrant for the determination of metals. Active research continues in the field of chelate chemistry. Some topics of current interest including the determination of the structure of chelate compounds and the synthesis of new chelating agents, especially those tailor-made to "fit" only certain metal ions and not others.

5–6. POLYNUCLEAR COMPLEXES

The complex ions we have discussed so far, those containing only a single central metal ion with its associated ligands, are called *mononuclear* complexes. It is possible to form complexes that are *polynuclear* and that contain two or more central metal ions per complex ion molecule. Some of the most common polynuclear complexes are those involving OH^- as a ligand. For example, Fe^{+++} forms a polynuclear complex by the reaction

$$2Fe^{+++} + 2OH^- \rightleftarrows Fe_2(OH)_2^{++++} \tag{5.45}$$

which also yields the K_{stab} expression

$$\frac{[Fe_2(OH)_2^{++++}]}{[Fe^{+++}]^2[OH^-]^2} = K_{stab} = 1 \times 10^{25} \tag{5.46}$$

Bismuth forms a polynuclear complex with OH^- which involves 6 bismuth ions and 12 hydroxide ions. The general reaction for the formation of a complex ion is therefore (neglecting ionic charges)

$$mM + nL \rightleftarrows M_mL_n \tag{5.47}$$

where M is the central metal ion and L is the ligand.

5–7. STABILITY VS. LABILITY

It is well to emphasize again the difference between the position of equilibrium of a reaction and the rate of a reaction, this time in connection with complex ions. We have seen that the terms *stability* and *instability* describe the extent of formation of a complex ion at equilibrium. A stable complex is one that has a large stability constant, and will occur, in the presence of excess ligand, predominantly in the form of the complex ion at equilibrium. The rate at which complexes form or decompose is described by the terms *labile* and *inert*. A labile complex ion reacts rapidly, while an inert complex ion reacts slowly. For example, calculations show that $Fe(CN)_6^{---}$ should be decomposed in strongly acidic solutions because the reaction

$$Fe(CN)_6^{---} + 6H^+ \rightleftarrows Fe^{+++} + 6HCN \tag{5.48}$$

lies to the right. However when $Fe(CN)_6^{---}$ is added to a strongly acidic solution, the complex ion species persists for a long time because the rate of (5.48) is slow. Chromium (III) complexes are also usually inert, and the reaction

$$Cr(H_2O)_6^{+++} + Y^{-4} \rightleftarrows CrY^- + 6H_2O \qquad (5.49)$$

(where Y^{-4} is the EDTA anion, and the ligand water is explicitly written in this case to stress the complex ion nature of the original species) takes a long time to go to completion, even though K_{stab} of CrY^- is about 10^{23}. On the other hand, the reaction

$$Ni(NH_3)_4^{++} + 4CN^- \rightleftarrows Ni(CN)_4^{--} + 4NH_3 \qquad (5.50)$$

occurs quite rapidly, so that $Ni(NH_3)_4^{++}$ is said to be a labile complex.

PROBLEMS

5.1. Calculate the concentrations of all of the species, ions and molecules, in the following solutions. (Neglect secondary equilibria, such as the basicity of the anions, in this problem.) All solutions have a total volume of 1.00 liter and contain:
- (a) 0.010 mole $AgNO_3$, 2.00 moles NH_3
- (b) 0.050 mole $Co(NO_3)_3$, 1.30 mole NH_3
- (c) 0.010 mole $Zn(NO_3)_2$, 1.00 mole ethylenediamine (en)
- (d) 0.10 mole $Cu(NO_3)_2$, 2.20 mole sodium tartrate
- (e) 0.010 mole $Fe(NO_3)_3$, 1.00 mole $Na_2C_2O_4$
- (f) 0.050 mole $CaCl_2$, 1.05 mole EDTA

5.2. Calculate (1) the final concentration of ligand, and (2) the total number of moles of complexing agent that must be added to dissolve completely the indicated precipitates in 1 liter of solution.
- (a) 0.010 mole AgCl in NH_3
- (b) 0.050 mole CaC_2O_4 in EDTA
- (c) 0.010 mole NiS in KCN
- (d) 0.020 mole $Al(OH)_3$ in KF
- (e) 0.010 mole Ag_2S in KCN

5.3. Calculate the molar solubility of the precipitate and the concentrations of all of the species in solutions made by mixing an excess of the solid precipitate with the given concentration of complexing agent (assume no volume change on mixing).
- (a) Ag_2S treated with 1.0 M NH_3
- (b) $Fe(OH)_3$ treated with 2.0 M $Na_2C_2O_4$
- (c) PbI_2 treated with 2.0 M sodium acetate
- (d) HgS treated with 0.10 M EDTA

5.4. Over what range of S^{--} concentration can the separation of zinc and cadmium be accomplished (by quantitative precipitation of one ion, but not the other) in a solution 0.010 M in both Zn^{++} and Cd^{++}

(a) without the addition of a complexing agent

(b) in a solution containing 1.0 M NH_3

(c) in a 2 M NaOH solution.

5.5. If NaOH is added to a solution containing Zn^{++}, first $Zn(OH)_2$ precipitates, then it redissolves with the formation of $Zn(OH)_4^{--}$. Calculate the concentrations of Zn^{++} and $Zn(OH)_4^{--}$ in equilibrium with solid $Zn(OH)_2$ at the following pH's: (a) 6.00 (b) 7.00 (c) 10.00 (d) 13.00 (e) 14.00.

5.6. A solution containing 0.010 M Ni^{++} and 0.010 M Co^{++} is made 1.0 M in NH_3.

(a) First, assuming no precipitation occurs, calculate the concentrations of Ni^{++}, $Ni(NH_3)_6^{++}$, Co^{++}, $Co(NH_3)_6^{++}$, and OH^-.

(b) Will either $Ni(OH)_2$ or $Co(OH)_2$ precipitate?

(c) Calculate the concentrations of the above species at equilibrium.

(d) Can this method be used as a means of separating Ni^{++} from Co^{++}?

5.7. A solution containing 0.010 M Ni^{++} and 0.010 M Zn^{++} is treated with 1.0 M ethylenediamine. Over what range of OH^- concentration will the quantitative separation of Ni^{++} from Zn^{++}, by precipitation of $Zn(OH)_2$ leaving $Ni(en)_2^{++}$ in solution, be successful? (The total concentration of Zn^{++} and $Zn(en)_2^{++}$ in solution is less than 0.1 percent of the original concentration of zinc.)

5.8. Calculate α_1 for HCN at pH's of 5.00, 7.00, and 11.00.

5.9. (a) Sketch an α_1 vs. pH curve for the acetate-acetic acid system (see Fig. 5.1).

(b) Calculate the concentration of uncomplexed Pb^{++} in a solution containing 1.0 mole of CH_3COOH and 0.010 mole of $Pb(NO_3)_2$ per liter of solution at pH's of 2.00, 4.00, and 8.00.

5.10. (a) Derive the following expression for the fraction of a diprotic acid, H_2X, totally ionized, α_2:

$$\alpha_2 = \frac{[X^{--}]}{C_T} = \frac{K_1K_2}{[H^+]^2 + K_1[H^+] + K_1K_2}$$

where $$C_T = [H_2X] + [HX^-] + [X^{--}]$$

(b) Sketch an α_2 vs. pH curve for tartaric acid.

(c) Calculate the concentration of uncomplexed Cu^{++} in a solution containing 0.10 mole of sodium tartarte and 0.0010 mole $Cu(NO_3)_2$ per liter of solution at pH's of 7.00, 5.00, and 3.00.

SUPPLEMENTARY READING

Bailar, J. C., Jr., ed., *The Chemistry of the Coordination Compounds*, New York: Reinhold, 1956.

Banks, J. E., Equilibria of Complex Formation, *J. Chem. Educ.*, **38**, 391 (1961).

Martell, A. E., The Behavior of Metal Complexes in Aqueous Solution, *J. Chem. Educ.*, **29**, 270 (1952).

Martell, A., and M. Calvin, *The Chemistry of Metal Chelate Compounds*, Englewood Cliffs, N.J.: Prentice-Hall, 1952.

Ringbom, A., Conditional Constants, *J. Chem. Educ.*, **35**, 282 (1958).

CHAPTER 6 OXIDATION-REDUCTION EQUILIBRIA

6–1. OXIDATION-REDUCTION REACTIONS

In all of the reactions discussed so far the oxidation state of the elements involved in the reactions remained constant. On the other hand, in oxidation-reduction (or *redox*) reactions, electrons are exchanged and oxidation states change. For example, in the reaction

$$Cu^{++} + Zn \rightleftarrows Cu + Zn^{++} \qquad (6.1)$$

the copper changes from a zero to a $+2$ oxidation state, while the zinc changes from a $+2$ to a zero state. Redox reactions can be separated into two *half-reactions*, an oxidation, or a loss of electrons, and a reduction, or a gain of electrons. For (6.1), the two half-reactions are

$$Zn - 2e \rightleftarrows Zn^{++} \quad (oxidation)$$
$$Cu^{++} + 2e \rightleftarrows Cu \quad (reduction)$$

A redox reaction may also be said to take place between an *oxidizing agent* and a *reducing agent*. An oxidizing agent is a substance which causes an oxidation to occur, while being reduced itself. A reducing agent is a substance causing a reduction to occur, being oxidized itself. In (6.1) zinc metal is the reducing agent and Cu^{++} is the oxidizing agent. Letting n' be the number of electrons gained or lost in the overall reaction, that is, the number of electrons exchanged between the oxidizing and reducing agents, for (6.1), $n' = 2$.

Another redox reaction is

$$5Fe^{++} + MnO_4^- + 8H^+ \rightleftarrows 5Fe^{+++} + Mn^{++} + 4H_2O \quad (6.2)$$

composed of the following half reactions:

$$5[Fe^{++} - e \rightleftarrows Fe^{+++}] \quad \text{(oxidation)}$$

$$MnO_4^- + 8H^+ + 5e \rightleftarrows Mn^{++} + 4H_2O \quad \text{(reduction)}$$

For (6.2), $n' = 5$. The separation of an oxidation-reduction reaction into two half reactions is a purely formal step, and is not the description of the actual *mechanism* of the reaction. Indeed, the mechanism of (6.2) is very complicated, and is said to occur in a series of steps involving manganese(III), (IV), and (V), and perhaps even higher oxidation states of iron, such as iron(IV)*. As long as we consider only equilibrium problems, we can divide a reaction in any way we choose, since all paths from products to reactants require the same net energy change and will yield the same value of the equilibrium constant. Many oxidation-reduction reactions occur by an exchange of a group which carries the electrons (e.g., a ligand), without an actual exchange of electrons themselves.

6–2. OXIDATION-REDUCTION POTENTIALS

HALF-REACTION POTENTIALS

How can we predict the extent of a redox reaction? There are no tables giving the equilibrium constants of specific redox reactions because the very large number of possible reactions makes a complete tabulation inconvenient. Instead characteristics of the half-reactions are tabulated and these are put in various combinations to determine equilibrium constants of the reactions of interest. While there are several possible quantities that can be used to characterize these half-reactions,

* H. Laitinen, *Chemical Analysis*, New York: McGraw-Hill, 1960, pp. 368–372.

the one conventionally used is the *potential* or *emf* of the half-reaction. We may recall that the potential (or voltage) in an electric circuit can be thought of as a driving force pushing or pulling the electrons through the wires of the circuit, just as the water pressure represents the force pushing water through the pipes of a plumbing system. Redox reactions are somewhat analogous to an electric circuit, in that electrons are transferred or flow from one substance to another. Every half-reaction has a *standard potential*, E^0, associated with it. For example, we may find in a table of standard potentials the following half-reactions:

$$Cu^{++} + 2e \rightleftarrows Cu \quad E^0 = 0.34 \text{ volts}$$
$$Zn^{++} + 2e \rightleftarrows Zn \quad E^0 = -0.76 \text{ volts}$$

POTENTIALS AND REDUCING ABILITY

The E^0 of a half-reaction is a measure of the tendency of the reaction to proceed to the right (relative to an arbitrarily chosen half-reaction). Suppose we wish to predict whether Zn will reduce Cu^{++} as in (6.1), or whether Cu will reduce Zn^{++} by the reaction

$$Cu + Zn^{++} \rightleftarrows Cu^{++} + Zn \tag{6.3}$$

Half-reactions with a large, positive E^0 tend to proceed to the right; the substance gaining electrons is a good oxidizing agent (i.e., a good "electron-puller"). Half-reactions with a large, negative E^0 do not tend to proceed to the right and the substance gaining electrons is a poor oxidizing agent (i.e., a poor "electron-puller"). The E^0 of a half-reaction, when written as a reduction (by convention*) is then a measure of "electron-pulling" power. Since the potential of the Cu^{++}, Cu couple, is more positive than that of the Zn^{++}, Zn couple, Cu^{++} is a better "electron-puller" than Zn^{++}; or, Cu^{++} is a better oxidizing agent than Zn^{++}, and so Cu^{++} will take electrons from Zn to form Zn^{++} and the reaction will proceed as in (6.1). A table of E^0's of some half-reactions is given in Appendix D. Some of the ideas discussed in the preceding paragraphs are illustrated in Figure 6.1.

REACTION POTENTIALS

These ideas can be made more quantitative, and we can predict not only if a reaction will proceed but to what extent it will proceed by

---

* Actually a chart of half-reactions may be written as either reductions or as oxidations. Although some authors still tabulate half-reactions as oxidations, there is some advantage in writing these as reductions (namely a better correspondence between the thermodynamic sign and the electrostatic sign in electrochemistry), and this practice, recommended by several international groups, is followed here.

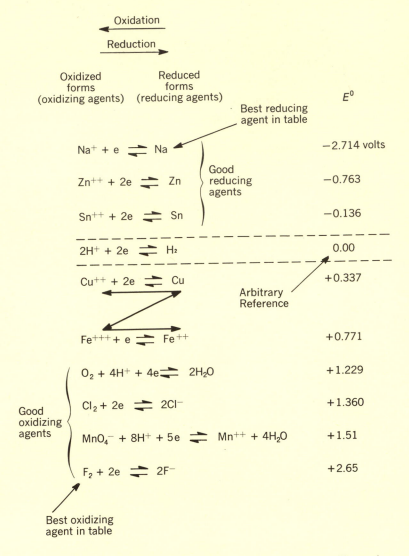

FIGURE 6.1. Selected standard electrode potentials illustrating reducing and oxidizing properties of substances in the half-reactions. Prediction about whether an oxidation-reduction reaction will occur can be made by the "Z-method." Join the two substances whose reaction is questioned (e.g., Cu and Fe^{+++}) with a line. Add lines indicating the directions of the half-reactions. If a figure "Z" is formed, the reaction is spontaneous (or $K > 1$) (e.g., Cu will react with Fe^{+++} to produce Fe^{++} and Cu^{++}).

calculating the E^0 for the total reaction, E^0_{rxn}. E^0_{rxn} is defined as the difference of the E^0's of the half-reactions, where the E^0 of the couple undergoing oxidation in the total reaction is subtracted from the E^0 of the couple undergoing reduction (that is, the couple subtracted is the one occurring in a direction opposite from that quoted in the table of standard potentials),

$$E^0_{rxn} = E^0_{red} - E^0_{oxid} \qquad (6.4)$$

For (6.1), E^0_{rxn} is given by

$$E^0_{rxn} = E^0_{Cu^{++},Cu} - E^0_{Zn^{++},Zn}$$

because in the reaction as written, Zn is getting oxidized. Therefore, for (6.1)

$$E^0_{rxn} = 0.34 - (-0.76) = +1.10 \text{ volts}$$

The positive value of E^0_{rxn} indicates that the reaction proceeds spontaneously as written, while the magnitude of E^0_{rxn} is a measure of the extent of the reaction. Consider the reaction as written in (6.3). Now the E^0 of the Cu^{++}, Cu couple is subtracted, since it is Cu which is now

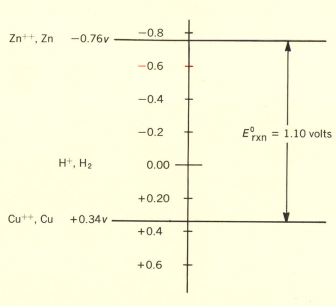

$$E^0_{rxn} = E^0_{Cu^{++},Cu} - E^0_{Zn^{++},Zn}$$

FIGURE 6.2. Calculating the E^0_{rxn} from the E^0's of the half-reactions.

getting oxidized. For the reaction as written in (6.3)

$$E^0_{rxn} = -0.76 - (0.34) = -1.10 \text{ volts}$$

The negative sign in E^0_{rxn} implies that the reaction proceeds spontaneously in the direction opposite to the way in which it is written (6.3). These calculations are illustrated in Figure 6.2.

Example 6.1. Which of these reactions will occur spontaneously?

$$2Fe^{+++} + 2I^- \rightleftarrows 2Fe^{++} + I_2 \tag{6.5}$$

$$2Fe^{++} + I_2 \rightleftarrows 2Fe^{+++} + 2I^- \tag{6.6}$$

That is, will Fe^{++} reduce I_2 or will Fe^{+++} oxidize I^-? What is E^0_{rxn} for the spontaneous reaction?

In the table of potentials in Appendix D we find the following entries:

$$Fe^{+++} + e \rightleftarrows Fe^{++} \quad E^0_{Fe^{+++},Fe^{++}} = 0.77 \text{ volts}$$

$$I_2 + 2e^{\cdot} \rightleftarrows 2I^- \quad E^0_{I_2, I^-} = 0.54 \text{ volts}$$

By inspection we see that Fe^{+++} is a better electron puller than I_2, or that Fe^{+++} is a better oxidizing agent than I_2, and therefore Fe^{+++} will oxidize I^- as in (6.5). For (6.5)

$$E^0_{rxn} = 0.77 - (0.54) = +0.23 \text{ volts}$$

Note that for (6.6)

$$E^0_{rxn} = 0.54 - (0.77) = -0.23 \text{ volts}$$

or the reaction is not spontaneous in the direction written.

In the calculation of E^0_{rxn} the number of electrons exchanged in the overall reaction, n', is not considered. This is because potential is analogous to pressure, and the number of electrons actually flowing is unrelated to the differences in "electron pressures" between the two couples.

6-3. EQUILIBRIUM CONSTANTS OF REDOX REACTIONS

The equilibrium constant for a redox reaction, K, can be calculated by the following equation:

$$\log K = \frac{n'E^0_{rxn}}{0.059} \tag{6.7}$$

at 25° C. The derivation of this formula is beyond the scope of this discussion and is given in most books on thermodynamics.* Using (6.7), the equilibrium constant of the reaction,

$$Cu^{++} + Zn \rightleftarrows Cu + Zn^{++}$$

can be calculated.

Since $E^0_{rxn} = 1.10$ volts, and $n' = 2$ for this reaction,

$$\log K = \frac{2(1.10)}{0.059} = 37.3 \tag{6.8}$$

$$K = 2 \times 10^{37} = \frac{[Zn^{++}]}{[Cu^{++}]} \tag{6.9}$$

Note that the equilibrium constant expression does not contain terms for solid Cu and Zn. Since these substances are solid, their activities are constant, and, by the same argument as that presented in Chapter 4, are taken as equal to one and included in the equilibrium constant. The K for the reverse reaction

$$Zn^{++} + Cu \rightleftarrows Cu^{++} + Zn$$

for which $E^0_{rxn} = -1.10$ volts, and $n' = 2$, is given by

$$\log K = \frac{2(-1.10)}{0.059} = -37.3 \tag{6.10}$$

$$K = 5 \times 10^{-38} = \frac{[Cu^{++}]}{[Zn^{++}]} \tag{6.11}$$

which is in agreement with (6.9).

Example 6.2. Calculate K for the reaction

$$2Fe^{+++} + 2I^- \rightleftarrows 2Fe^{++} + I_2$$

From Example 6.1, $E^0_{rxn} = 0.23$ volts, $n' = 2$.

$$\log K = \frac{2(0.23)}{0.059} = 7.8$$

$$K = 6 \times 10^7 = \frac{[Fe^{++}]^2[I_2]}{[Fe^{+++}]^2[I^-]^2} \tag{6.12}$$

---

* W. J. Moore, *Physical Chemistry*, Englewood Cliffs, N.J.: Prentice-Hall, 1962, p. 388, and L. K. Nash, *Elements of Chemical Thermodynamics*, Reading, Mass.: Addison-Wesley, 1962, p. 80.

Example 6.3.   Calculate K for the reaction

$$5Fe^{++} + MnO_4^- + 8H^+ \rightleftharpoons 5Fe^{+++} + Mn^{++} + 4H_2O$$

For the reaction as written, $E^0{}_{rxn} = 0.74$ volts, $n' = 5$.

$$\log K = \frac{5(0.74)}{0.059} = 62.7$$

$$K = 5 \times 10^{62} = \frac{[Fe^{+++}]^5[Mn^{++}]}{[Fe^{++}]^5[MnO_4^-][H^+]^8}$$

6–4. OXIDATION-REDUCTION PROBLEMS

Once the equilibrium constant for a redox reaction is determined, and the equilibrium constant expression written, problems of the type considered in the previous chapters can be solved. Often these problems involve pH, solubility, and complex ion effects.

Example 6.4.   One liter of a 0.10 M Cu^{++} solution is shaken with metallic Zn. Calculate $[Zn^{++}]$ and $[Cu^{++}]$ in this solution at equilibrium.

For this system, at equilibrium, from (6.9)

$$K = 2 \times 10^{37} = \frac{[Zn^{++}]}{[Cu^{++}]}$$

Since K is very large, almost all of the Cu^{++} will be used up and an equivalent amount of Zn^{++} formed, so that

$$[Zn^{++}] \approx 0.10 \ M$$

$$[Cu^{++}] = \frac{0.10}{2 \times 10^{37}} = 5 \times 10^{-39} \ M *$$

Example 6.5.   At what pH's will the reaction of 0.10 mole of I_3^- ** and 0.10 mole of $HAsO_2$ in 1.0 liter of solution, according to the reaction

$$I_3^- + HAsO_2 + 2H_2O \rightleftharpoons H_3AsO_4 + 3I^- + 2H^+$$

be quantitative?

From the potentials of the two half-reactions,

$$E^0{}_{rxn} = E^0{}_{I_3^-,I^-} - E^0{}_{HAsO_2,H_3AsO_4} = 0.536 - 0.559$$

---

* Note that this quantity is really such a small number that it has no physical meaning. A concentration of 5×10^{-39} moles/liter is equivalent to 3×10^{-15} ions/liter, one ion per 10^{14} liters. In reality, equilibrium considerations, indeed all thermodynamic considerations, only apply when a fairly large number of ions or molecules are considered.

** I_2 in the presence of I^- forms the triiodide complex, I_3^-, $I_2 + I^- \rightleftharpoons I_3^-$, $K = 710$. The E^0 for the I_3^-, I^- couple is just about the same as that of the I_2(solid), I^- couple.

$E^0_{rxn} = -0.023$ volts and $n' = 2$

$$\log K = \frac{2(-0.023)}{0.059} = -0.78$$

$$K = 0.16 = \frac{[H_3AsO_4][I^-]^3[H^+]^2}{[HAsO_2][I_3^-]}$$

For a quantitative reaction, at least 99.9 percent of the I_3^- and $HAsO_2$ must react, so that

$$[I_3^-] = 10^{-3} \times 0.10 = 10^{-4} M \quad [HAsO_2] = 10^{-3} \times 0.10 = 10^{-4} M$$

$$[H_3AsO_4] = 0.10 - 10^{-4} \approx 0.10 M$$

$$[I^-] = 3(0.10 - 10^{-4}) \approx 0.30 M$$

Introducing these quantities into the equilibrium constant expression

$$[H^+]^2 = \frac{0.16 \times 10^{-8}}{2.7 \times 10^{-3}} = 59 \times 10^{-8}$$

$$[H^+] = 7.7 \times 10^{-4} M \quad \text{or} \quad pH = 3.11$$

Since the reaction becomes more complete with decreasing $[H^+]$ (Le Chatelier's Principle), any pH larger than 3.11 will yield a quantitative reaction.

Example 6.6. When a solution containing $2.0 \times 10^{-2} M$ Cu^{++} is made $1.0 M$ in I^- (by the addition of KI), I_3^- is formed and CuI precipitates. Calculate the $[Cu^{++}]$ and the $[I_3^-]$ in the resulting solution.

The overall reaction

$$2Cu^{++} + 5I^- \rightleftarrows 2CuI + I_3^-$$

can be considered in two steps; the redox reaction

$$2Cu^{++} + 3I^- \rightleftarrows 2Cu^+ + I_3^-$$

and the precipitation reaction

$$2Cu^+ + 2I^- \rightleftarrows 2CuI$$

The equilibrium constants for these reactions are

$$\frac{[Cu^+]^2[I_3^-]}{[Cu^{++}]^2[I^-]^3} = 5 \times 10^{-13}$$

$$[Cu^+][I^-] = K_{sp} = 1.4 \times 10^{-12}$$

Introducing the $[Cu^+]$ term of the K_{sp} expression into the redox expression

$$\frac{[I_3^-]}{[Cu^{++}]^2[I^-]^5} = \frac{5 \times 10^{-13}}{(1.4 \times 10^{-12})^2} = 2.5 \times 10^{11}$$

The magnitude of this number indicates that the reaction between Cu^{++} and I^- is quite complete under these conditions, so that

$$[I_3^-] \approx \tfrac{1}{2}(2.0 \times 10^{-2}) = 1.0 \times 10^{-2} \; M$$

$$[I^-] = 1.0 - 3(1.0 \times 10^{-2}) \approx 1.0 \; M$$

$$[Cu^{++}]^2 = \frac{1.0 \times 10^{-2}}{(1.0)^3 \times 2.5 \times 10^{11}} = 5 \times 10^{-14}$$

$$[Cu^{++}] = 2.2 \times 10^{-7} \; M$$

Note that this reaction would not occur if CuI did not precipitate.

6–5. RATE OF OXIDATION-REDUCTION REACTIONS

We must again use caution in predicting whether reactions will actually occur from equilibrium considerations alone. For example, the K for the reaction

$$Sn^{++} + 2Fe^{+++} \rightleftarrows Sn^{++++} + 2Fe^{++} \tag{6.12}$$

in noncomplexing media is about 10^{21}. Yet when Fe^{+++} and Sn^{++} are mixed, they react very slowly and attain equilibrium only after a very long time. On the other hand, the reaction

$$Sn^{++} + I_2 \rightleftarrows Sn^{++++} + 2I^- \tag{6.13}$$

for which K is 10^{14}, is very rapid. How can we account for the difference in the rate with which Sn^{++} reacts with Fe^{+++} and I_2? Some investigators explain that (6.12) is slow in noncomplexing media because it involves the collision of two positively charged particles (which tend to repel each other). Others point out that (6.12) apparently involves the simultaneous collision of three particles, a relatively improbable occurrence, to cause the exchange of two electrons at once for oxidizing tin(II) to tin(IV) (assuming an intermediate oxidation state of tin is not possible). Truthfully our knowledge of the mechanisms of redox reactions is not well enough established for us to generally predict which reactions will be rapid and which will not.

It is fortunate that some redox reactions are slow. Equilibrium calculations indicate that almost all large organic molecules (including those which make up you and I) should spontaneously decompose into simpler substances, such as methane (CH_4) and water. These decomposition reactions are exceedingly slow because of the large amount of (activation) energy needed to break the strong bonds in the original molecules. Similarly, oxygen is a very good oxidizing agent (E^0 for the O_2—H_2O couple is $+1.23$ volts), and except for the strength of the oxygen-to-oxygen bond, leading to slow oxidations by molecular oxygen, many substances would react with oxygen and decompose.

PROBLEMS

6.1. Calculate E^0_{rxn} and K and write the equilibrium constant expressions for the following reactions (as written)

(a) $2Cr^{++} + Sn^{++++} \rightleftarrows 2Cr^{+++} + Sn^{++}$

(b) $Br_2 + 2Fe^{++} \rightleftarrows 2Br^- + 2Fe^{+++}$

(c) $Cr_2O_7^{--} + 3H_2O_2 + 8H^+ \rightleftarrows 2Cr^{+++} + 3O_2 + 7H_2O$

(d) $Cl_2 + 2I^- \rightleftarrows 2Cl^- + I_2$

(e) $2AgCl + 2Hg \rightleftarrows 2Ag + Hg_2Cl_2$

6.2. Calculate (1) the K for the reaction and (2) the concentrations of the ions at equilibrium when the following reactants are mixed. An excess of the metallic reductant is added in every case.

(a) $0.0200\ M$ $Cu(NO_3)_2$ and Zn

(b) $0.0500\ M$ $SnCl_4$ and Ni

(c) $0.100\ M$ $Cu(NO_3)_2$ and Ag

(d) $0.0100\ M$ $AgNO_3$ and Pb

6.3. At what pH's, if any, will solid PbO_2 quantitatively oxidize a $0.0100\ M$ solution of Mn^{++} to MnO_4^-?

6.4. When an element can exist in more than two oxidation states one of the states may be unstable with respect to the others. For example, copper can exist as Cu, Cu^+, and Cu^{++}. In a noncomplexing medium Cu^+ is unstable and *disproportionates* into Cu and Cu^{++} by the reaction

$$2Cu^+ \rightleftarrows Cu^{++} + Cu$$

(a) Calculate the K for this disproportionation reaction.

(b) If 0.010 mole of a Cu^+ salt is diluted to 1.0 liter with water, what are the final concentrations of Cu^+ and Cu^{++} in the solution at equilibrium?

(c) Repeat the calculations of part (b) for $Cu(NH_3)_4^{++}$ and $Cu(NH_3)_2^+$ in a solution $1\ M$ in NH_3.

6.5. Which of the following species disproportionate to a large extent in noncomplexing media? If the species disproportionates, calculate the K of the reaction.

(a) Hg_2^{++} (c) Au^+
(b) Sn^{++} (d) I_2 (to I^- and IO^-)

6.6. An analyst wants to reduce a $0.010\ M\ U^{++++}$ solution to U^{+++} with cadmium metal, by the reaction

$$Cd + 2U^{++++} \rightleftarrows Cd^{++} + 2U^{+++}$$

(a) Calculate K for this reaction.
(b) Is quantitative reduction possible?
(c) The analyst decides that perhaps the addition of CN^-, complexing Cd^{++} to $Cd(CN)_4^{--}$, will make the reaction more complete. Calculate the $[CN^-]$ needed for the quantitative reduction of U^{++++}. (Assume U^{++++} and U^{+++} are not complexed by CN^-.)

6.7. Cobalt(III) can be produced by oxidation of Cobalt(II) in an ammoniacal solution with H_2O_2 by the reaction

$$2Co(NH_3)_6^{++} + H_2O_2 + 2NH_4^+ \rightleftarrows 2Co(NH_3)_6^{+++} + 2NH_3 + 2H_2O$$

Calculate the ratio Cobalt(III)/Cobalt(II) in a solution containing $0.10\ M\ H_2O_2$ in excess, in a $1\ M\ NH_3$—$1\ M\ NH_4^+$ buffer.

6.8. The equilibrium constant for the reaction

$$AuCl_4^- + 2Au + 2Cl^- \rightleftarrows 3AuCl_2^-$$

was obtained by analyzing a solution for $AuCl_4^-$. A solution initially containing $5.57 \times 10^{-4}\ M\ AuCl_4^-$ in $2.06\ M$ HCl, was found to contain $4.67 \times 10^{-4}\ M\ AuCl_4^-$ when allowed to come to equilibrium with an excess of solid Au. Calculate K for the reaction.

CHAPTER 7 ACTIVITY AND ACTIVITY COEFFICIENTS

7–1. ACTIVITY VS. CONCENTRATION

The theory of equilibrium developed in the previous chapters was based upon the assumption that we were dealing with *ideal* solutions. An ideal solution is one in which there are no interionic forces and in which the ions are distributed at random throughout the solution. In an ideal solution the equilibrium constant expression can be written in terms of the concentrations of the various species and the equilibrium constant is truly a constant, independent of the concentrations of the various ions in the particular solution. Suppose we want to test the constancy of this concentration equilibrium constant in a real (nonideal) solution. Let us measure the solubility of AgCl in solutions containing varying amounts of KNO_3. Since potassium ions and nitrate ions do not react with silver ions or chloride ions (that is, KNO_3 is an *inert electrolyte* as far as AgCl is concerned), we would predict that the solubility of AgCl would be independent of the concentration of KNO_3

and would be governed by the expression

$$[Ag^+][Cl^-] = K_{sp} \tag{7.1}$$

The results that are actually obtained for this experiment are shown in Table 7.1 (the symbol μ, shown in the table, will be discussed later). The

TABLE 7.1.   Solubility of AgCl in Solutions Containing Various Amounts of KNO_3

KNO_3 conc. M	Solubility of AgCl (s) (moles/liter)	μ	$\sqrt{\mu}$	$\log s^2$
0.000	1.28×10^{-5}	1.28×10^{-5}	3.6×10^{-3}	-9.786
0.001	1.32×10^{-5}	0.00101	3.2×10^{-2}	-9.758
0.005	1.38×10^{-5}	0.005	7.1×10^{-2}	-9.720
0.010	1.43×10^{-5}	0.010	0.10	-9.690

Data of S. Popoff and E. W. Neuman, *J. Phys. Chem.* **34**, 1853 (1930).

experiment shows, contrary to our expectations, that the solubility of AgCl *increases* with increasing concentrations of KNO_3.

If we write our equilibrium constant expressions in terms of *activities* rather than concentrations, we can maintain the constancy of the equilibrium constant and allow for the effect of the inert electrolyte. We can regard the activity as an " effective concentration"; in an ideal solution the concentration equals the activity. Instead of using concentrations in equilibrium constant expressions, we will use activities. Now (7.1) would be given by

$$a_{Ag^+} a_{Cl^-} = \mathbf{K}_{sp} \tag{7.2}$$

where a_{Ag^+} is the activity of silver ion, a_{Cl^-} is the activity of chloride ion, and $\mathbf{K}_{sp}$ is the true (thermodynamic) solubility product for AgCl. The activity of an ion X, a_X, is related to its concentration, $[X]$, by the equation

$$a_X = f_X[X] \tag{7.3}$$

where f_X is the *activity coefficient*. As solutions become more dilute (contain lower concentrations of ions) f_X approaches 1. The activity coefficient is 1 in ideal (infinitely dilute) solutions and varies with the total ionic concentration of all species in the solution.

We can now use the concept of activity to define a truly constant equilibrium constant. From (7.2) and (7.3) we write

$$f_{Ag^+}[Ag^+]f_{Cl^-}[Cl^-] = \mathbf{K}_{sp} \tag{7.4}$$

$$[Ag^+][Cl^-] = \frac{\mathbf{K}_{sp}}{f_{Ag^+} \, f_{Cl^-}} = K_{sp} \qquad (7.5)$$

where $\mathbf{K}_{sp}$ is a true constant, as opposed to K_{sp}, which varies with the concentration of KNO_3. If f_{Ag^+} and f_{Cl^-} decrease with increasing ionic concentrations, then K_{sp}, the concentration equilibrium constant, will increase, so that the solubility of $AgCl$ will increase.

7–2. IONIC STRENGTH AND ACTIVITY COEFFICIENTS

IONIC STRENGTH

The activity coefficient varies with the total ionic population of the solution. A useful measure of the total ionic population of a solution is the *ionic strength*, μ. The ionic strength of a solution containing ion A, with a charge of Z_A, ion B, with a charge of Z_B, etc., is defined by the equation

$$\mu = \tfrac{1}{2}[A]Z_A^2 + \tfrac{1}{2}[B]Z_B^2 + \cdots \qquad (7.6)$$

or in a general form

$$\mu = \tfrac{1}{2} \sum_i [i]Z_i^2 \qquad (7.7)$$

Example 7.1. Calculate the ionic strength of a solution of 0.10 M $BaCl_2$.

In this solution: $[Ba^{++}] = 0.10 \ M$ $[Cl^-] = 0.20 \ M$
$$Z_{Ba^{++}} = 2 \quad Z_{Cl^-} = -1$$
$$\mu = \tfrac{1}{2}(0.10)(2)^2 + \tfrac{1}{2}(0.20)(-1)^2 = 0.30 \quad \textit{Answer}$$

Example 7.2. Calculate the ionic strength of a solution containing 0.10 M $MgCl_2$ and 0.20 M $Al_2(SO_4)_3$.

It is sometimes convenient to calculate μ by constructing a table of all of the ionic species in the solution.

Species, i	$[i]$	Z_i	Z_i^2	$[i]Z_i^2$
Mg^{++}	0.10	2	4	0.40
Cl^-	0.20	-1	1	0.20
Al^{+++}	0.40	3	9	3.60
SO_4^{--}	0.60	-2	4	2.40

$$\sum_i [i]Z_i^2 = 6.60$$

$$\mu = \tfrac{1}{2}\sum_i [i]Z_i^2 = \tfrac{1}{2}(6.60) = 3.30 \quad \textit{Answer}$$

MEASUREMENT OF ACTIVITY COEFFICIENTS

Activity coefficients can be determined experimentally by several different methods. The activity coefficient of a single ion, such as f_{Ag^-} or f_{Cl^-}, is not measureable because the concentration of a single ion cannot be varied without varying the concentration of an oppositely charged ion (electroneutrality principle). It is therefore usual to discuss the *mean activity coefficient* of a substance, $f_\pm$, which is related to the individual ionic activity coefficients of the substance $M_m N_n$ by

$$f_{M_m N_n} = f_M{}^m f_N{}^n = f_\pm{}^{m+n} \qquad (7.8)$$

For example,

$$f_{NaCl} = f_{Na^+} f_{Cl^-} = f_\pm{}^2$$

$$f_{BaCl_2} = f_{Ba^{++}} f_{Cl^-}{}^2 = f_\pm{}^3$$

Activity coefficients can be determined by measuring the solubility of

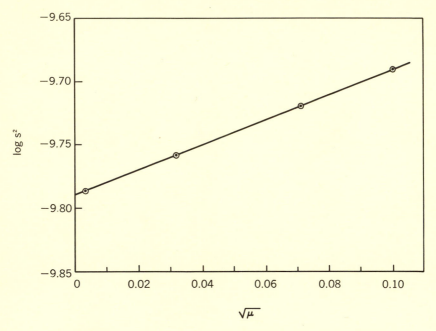

FIGURE 7.1.   Variation of the solubility of AgCl, s, with ionic strength, μ, in solutions containing different amounts of KNO_3.
$\log s^2 = \log K_{sp}$ at any μ $\log s^2 = \log K_{sp}$ at $\mu = 0$.

a precipitate in solutions of different ionic strengths. The determination of the solubility of AgCl in solutions of different KNO_3 concentrations, described in (7.1), is just such an experiment. Rewriting (7.5) in terms of mean activity coefficients

$$[Ag^+][Cl^-] = \frac{K_{sp}}{f_\pm^2} = s^2 = K_{sp} \tag{7.9}$$

where s is the molar solubility of AgCl. At any given ionic strength μ, calculated from the concentrations of K^+, NO_3^-, Ag^+, and Cl^- in the solution, we can determine K_{sp} (that is, s^2) (Table 7.1). When $\log K_{sp}$ is plotted against $\sqrt{\mu}$, a straight line results (Fig. 7.1). By extrapolation of this line to $\mu = 0$, where $f_\pm = 1$, we can determine the thermodynamic solubility product, K_{sp}. Once we have found this, we can calculate $f_\pm$ at any ionic strength, since from (7.9)

$$f_\pm = \sqrt{K_{sp}/K_{sp}} \tag{7.10}$$

Similar techniques based on the variation of other equilibrium constants, such as K_a and K_{stab}, with ionic strength, can also be used to find the thermodynamic equilibrium constants, K_a and K_{stab}. These values can then be used for the calculation of activity coefficients.

In general the activity coefficient may be thought of as an adjustable

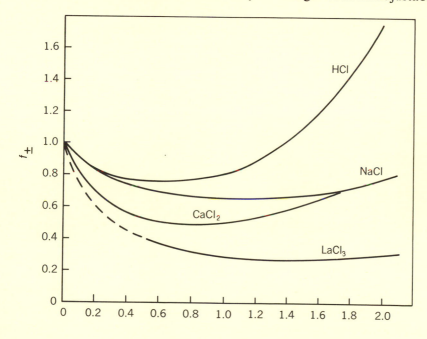

FIGURE 7.2. Mean activity coefficients, $f_\pm$, of several salts as a function of ionic strength.

parameter (or in student parlance, a "fudge factor") which can be used to correct a varying equilibrium "constant." Once these activity coefficients have been measured and tabulated, we can use them to adjust a thermodynamic equilibrium constant to the particular experimental conditions of interest. Experimentally measured activity coefficients of several substances at different ionic strengths are shown in Fig. 7.2.

7–3. THEORETICAL CALCULATION OF ACTIVITY COEFFICIENTS

THEORY OF ACTIVITY COEFFICIENTS

Let us consider why the activity (or effective concentration) of an ion varies with ionic strength. Actually a number of factors may contribute to the activity coefficient, including electrical effects, formation of unsuspected complex ions or undissociated species, and changes in the electrical insulating property (the *dielectric constant*) of the solvent. Contribution of electrical effects to the variation of the activity coefficient with ionic strength can be treated theoretically. These electrical effects are probably the major cause of deviation of the activity from the actual concentration in dilute solutions.

Consider a very dilute solution containing only silver and chloride ions. These ions are essentially "bare" (except for the ever present sheath of water molecules around each), are distributed randomly, and are free to move about unhindered. If an electrolyte, such as KNO_3, is added to the solution, the total ionic population of the solution will increase and the ions will be much closer together. The ions will not now be distributed randomly through the solution, because of *electrostatic forces* (i.e., forces arising from the attraction of oppositely charged species). The positive silver ion will attract negatively charged nitrate ions around it, while the negative chloride ion will attract potassium cations (Fig. 7.3), so that the silver and chloride ions will have *ionic atmospheres* around them. From a thermodynamic viewpoint this electrostatic interaction produces an extra free energy term (that is, in addition to the free energy of the reaction in the absence of electrostatic effects) and this additional energy is related to the activity coefficient. From a kinetic viewpoint the ionic atmospheres can be thought of as retarding the motion of the ions. As the ionic strength increases, the ionic atmospheres become more closely packed and the activity coefficient decreases. At still higher ionic concentrations the ions affect the solvent itself, decreasing its activity coefficient and thereby *increasing* the activity coefficients of the solute species (see Fig. 7.2).

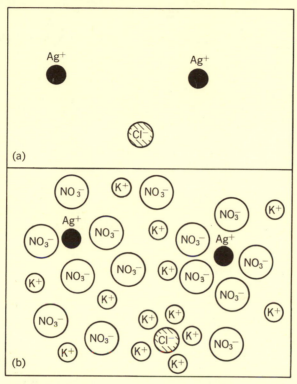

FIGURE 7.3. Schematic diagram of silver and chloride ions in solutions of (A) low ionic strength and (B) high ionic strength (with added KNO_3). Ionic sizes are approximate and water molecules are not indicated.

THE DEBYE-HÜCKEL THEORY

Using this model of ions with ionic atmospheres, and calculating the variation of the energy of the ions with ionic strength, Debye and Hückel were able to treat the effect of electrostatic forces on the activity coefficient quantitatively. Although the derivation of their now well known equation is rather difficult, the result for dilute solutions is quite simple.* The activity coefficient, f_i, of a single ion, i, of charge Z_i, is related to

* For a clear discussion of the derivation of the Debye-Hückel equation, see W. J. Moore, *Physical Chemistry*, 3rd. ed., Englewood Cliffs, N.J.: Prentice Hall, 1962, pp. 351–357.

the ionic strength μ, at 25° C in water, by the expression

$$\log f_i = \frac{-0.5\, Z_i^2\sqrt{\mu}}{1+\sqrt{\mu}} \tag{7.11}$$

The mean activity coefficient of an electrolyte M_mN_n is similarly given by

$$\log f_\pm = \frac{-0.5\, Z_M Z_N \sqrt{\mu}}{1+\sqrt{\mu}} \tag{7.12}$$

where Z_M is the charge of M and Z_N is the charge of N (without regard to sign).

Example 7.3.   Calculate the mean activity coefficient in 0.04 M HCl.

$$\mu = \tfrac{1}{2}(1)^2(0.04) + \tfrac{1}{2}(-1)^2(0.04) = 0.04$$

$$\sqrt{\mu} = 0.20 \quad Z_M = Z_N = 1$$

$$\log f_\pm = \frac{-0.5\,(1)(1)\sqrt{\mu}}{1+\sqrt{\mu}} = \frac{-0.5(0.2)}{1.2} = -0.0833$$

$$\log f_\pm = -1.00 + 0.9167$$

$$f_\pm = 0.825 \quad \textit{Answer}$$

The experimentally determined value of $f_\pm$ for HCl at an ionic strength of 0.04 is 0.843.

Example 7.4.   Calculate the mean activity coefficient in 0.01 M CaCl$_2$.

$$\mu = \tfrac{1}{2}(2)^2(0.01) + \tfrac{1}{2}(-1)^2(0.02) = 0.03$$

$$\sqrt{\mu} = 0.173 \quad Z_M = 2 \quad Z_N = 1$$

$$\log f_\pm = \frac{-0.5\,(2)(1)(0.173)}{1.173} = -0.1475$$

$$f_\pm = 0.712 \quad \textit{Answer}$$

The experimentally determined value of $f_\pm$ in 0.01 M CaCl$_2$ is 0.725.

Reference to (7.12) allows some qualitative statements about activity coefficients to be given. Uncharged molecules ($Z = 0$) behave most ideally, while the activity coefficients of ions decrease with increasing ionic charge. The derivation of the Debye-Hückel equation assumes very low ionic strengths, so that calculated activity coefficients are closest to experimental ones in very dilute solutions. Calculated and experimentally determined values of the activity coefficients for HCl and CaCl$_2$ at various concentrations are shown in Table 7.2.

TABLE 7.2. Calculated and Experimental Values of Activity Coefficients of HCl and $CaCl_2$ at Various Concentrations

Conc	f_{HCl}		f_{CaCl_2}	
M	Calc.	Exper.	Calc.	Exper.
0.001	0.966	0.965	0.89	0.89
0.002	0.952.	0.952	0.85	0.85
0.005	0.927	0.928	0.76	0.78
0.01	0.901	0.904	0.71	0.72
0.02	0.868	0.875	0.64	0.66
0.05	0.809	0.830	0.53	0.57
0.1	0.759	0.796	0.44	0.51
0.2	0.700	0.767	0.37	0.48
0.5	0.62	0.76	0.28	0.52
1.0	0.56	0.81	0.23	0.71
2.0	0.51	1.01		
3.0	0.48	1.32		

Clearly the calculated values of the activity coefficient differ quite significantly from experimental values at concentrations above about 0.05 *M*. Unfortunately, in actual experiments, we are usually interested in solutions much more concentrated, so that the Debye-Hückel equation may not be useful for many practical applications. Various attempts have been made to extend this treatment to more concentrated solutions by taking account of other factors that might influence the activity coefficients. Although some progress has been made in these treatments, most extensions contain "adjustable parameters" which cannot be calculated without recourse to experiment. In extremely dilute solutions ($\sqrt{\mu} \ll 0.05$), the Debye-Hückel equation becomes

$$\log f_{\pm} \approx -0.5\, Z_M Z_N \sqrt{\mu} \qquad (7.13)$$

Therefore, when determining thermodynamic equilibrium constants by extrapolation of experimental data in dilute solution, a plot of log K vs. $\sqrt{\mu}$ yields a straight line at low ionic strengths (Fig. 7.1).

7–4. ACTIVITY COEFFICIENTS AND EQUILIBRIUM PROBLEMS

How different will concentrations calculated from equilibrium constant expressions be when activity coefficients are considered as compared to concentrations calculated neglecting these effects (i.e., letting $f_{\pm} = 1$)? Consider the problem of finding [H$^+$] in a solution containing 0.010 M HAc and 0.10 M KCl. The thermodynamic ioniza-tion constant for HAc, $\mathbf{K}_a$, is 1.754×10^{-5}. Assuming that concentra-tions are the same as activities, we obtain

$$\mathbf{K}_a = 1.754 \times 10^{-5} = \frac{a_{H^+} a_{Ac^-}}{a_{HAc}} \approx \frac{[H^+][Ac^-]}{[HAc]} \qquad (7.14)$$

Solving this in the usual way, using the quadratic expression, we obtain $[H^+] = 4.10 \times 10^{-4}\ M$. This type of calculation, made in the previous chapters, is often sufficient for an estimation of the concentration. To take the presence of KCl into consideration, we must correct the thermo-dynamic equilibrium constant, $\mathbf{K}_a$, for the activity coefficients which are not 1, and calculate the concentration equilibrium constant, K_a,

$$\mathbf{K}_a = \frac{a_{H^+} a_{Ac^-}}{a_{HAc}} = \frac{f_{H^+}[H^+] f_{Ac^-}[Ac^-]}{f_{HAc}[HAc]} = \frac{f_{H^+} f_{Ac^-}}{f_{HAc}} K_a \qquad (7.15)$$

$$\log K_a = \log \mathbf{K}_a + \log f_{HAc} - \log f_{H^+} - \log f_{Ac^-} \qquad (7.16)$$

Applying the Debye-Hückel equation, (7.11), for the calculation of the activity coefficients, we obtain

$$\log f_{HAc} = 0 \qquad (7.17)$$

$$\log f_{H^+} = \log f_{Ac^-} = \frac{-0.5\sqrt{\mu}}{1 + \sqrt{\mu}} \qquad (7.18)$$

Combining (7.16), (7.17), and (7.18),

$$\log K_a = \log \mathbf{K}_a + \frac{\sqrt{\mu}}{1 + \sqrt{\mu}} \qquad (7.19)$$

For 0.10 M KCl, $\mu = 0.10$ (neglecting the slight contribution from the ionization of the HAc), so that K_a is 2.91×10^{-5}. Now we calculate a $[H^+]$ of $5.25 \times 10^{-4}\ M$ (a value 25 percent larger than that obtained in the first calculation). Actually we have seen that the Debye-Hückel equation tends to overcorrect at higher ionic strengths. A more exact calculation results when experimental activity coefficients, or an experi-mental K_a, can be found in the literature which have been determined for the solution of interest. In this particular case, K_a for HAc in 0.10 M KCl has been measured as 2.843×10^{-5}, yielding a $[H^+]$ of $5.19 \times 10^{-4}\ M$, a value slightly smaller than that calculated by the Debye-Hückel equation.

Example 7.5. Calculate the solubility of AgCl in a 0.010 M KNO_3 solution, without neglecting activity coefficient corrections, if $\mathbf{K}_{sp} = 1.64 \times 10^{-10}$.

$\mu = 0.010$ (neglecting contributions from Ag^+ and Cl^-)

$\sqrt{\mu} = 0.10$

$K_{sp} = a_{Ag^+}a_{Cl^-} = [Ag^+][Cl^-]f_{Ag^+}f_{Cl^-} = K_{sp}f_{Ag^+}f_{Cl^-}$

$\quad \log K_{sp} = \log K_{sp} - \log f_{Ag^+} - \log_{Cl^-}$

$\quad \log f_{Ag^+} = \log f_{Cl^-} = \dfrac{-0.5\sqrt{\mu}}{1+\sqrt{\mu}}$

$\quad \log K_{sp} = \log K_{sp} + \dfrac{\sqrt{\mu}}{1+\sqrt{\mu}}$

$\quad \log K_{sp} = -9.786 + 0.091 = -9.695 = -10 + 0.305$

$\qquad K_{sp} = 2.02 \times 10^{-10} \quad = s^2$

$\qquad\quad s = 1.42 \times 10^{-5}$ moles/liter   *Answer*

Compare this answer with the experimental result (Table 7.1) and with the solubility calculated neglecting activity coefficients.

7–5. CONCLUSION

This discussion of activity corrections and our inability to calculate, in advance and with a high degree of accuracy, activity coefficients in moderately concentrated solutions, may be discouraging. But this does not mean that our equilibrium theory is not useful for practical calculations. As long as we bear in mind that the calculations are only approximations and may not represent *exactly* the conditions in a real solution, the method is fruitful. We must then include activity corrections along with slow rates of reactions among the possible effects which limit the applicability and accuracy of equilibrium calculations. Another effect that limits the accuracy of equilibrium calculations is the presence of unsuspected equilibria or ion pair formation. For example, suppose we want to predict the solubility of AgCl in 0.1 M KCl. If we calculated the solubility and considered only activity effects (and, of course, the common ion effect), the result would be significantly in error. We must also consider the formation (perhaps unsuspected) of the complex ion $AgCl_2^-$, which leads to much higher solubility of AgCl than that predicted by activity effects alone. Often serious deviations between predictions based on theoretical calculation and experiment will uncover new reactions or effects and lead to a better understanding of the chemistry of the system.

PROBLEMS

7.1. Calculate the ionic strengths of the following solutions.
 (a) 0.10 M $AlCl_3$
 (b) 0.050 M $K_4Fe(CN)_6$
 (c) 0.050 M K_2SO_4 and 0.020 M $Al_2(SO_4)_3$
 (d) 0.10 M NaCl and 0.10 M Na_2SO_4

7.2. Calculate, using the Debye-Hückel equation, the mean activity coefficient of HCl and $MgCl_2$ in each solution in Problem 7.1.

7.3. From the data in Table 7.1 and Figure 7.1, calculate the thermodynamic solubility product, K_{sp}, of AgCl, and the mean activity coefficient of AgCl at each concentration of KNO_3.

7.4. Calculate the solubility of CuBr ($K_{sp} = 5.3 \times 10^{-9}$) in solutions of ionic strength 0.0, 0.010, and 0.10.

7.5. Calculate the $[H^+]$ in a solution of 0.10 M HAc and 0.10 M NaAc, (a) neglecting activity corrections (i.e., assuming $f_\pm = 1$), (b) calculating the ionic strength of the solution and computing K_a from $\mathbf{K}_a$ ($\mathbf{K}_a = 1.754 \times 10^{-5}$).

7.6. Write equations representing the concentration equilibrium constant, K, in terms of the thermodynamic equilibrium constant, $\mathbf{K}$, and the ionic strength, μ, for the following: (a) K_{sp} $Fe(OH)_3$, (b) K_{sp} $PbBr_2$, (c) K_a HF, (d) K_1 and K_2 H_2CO_3, (e) K_{stab} $Cd(NH_3)_4^{++}$.

SUPPLEMENTARY READING

DeFord, D., The Reliability of Calculations Based on the Law of Chemical Equilibrium, *J. Chem, Educ.*, **31**, 460 (1954). Discusses activity corrections and other effects which limit the accuracy of equilibrium calculations.

Hart, C. S., The Use of Activities in Student Calculations, *J. Chem. Educ.*, **32**, 314 (1955).

Laitinen, H. A., *Chemical Analysis*, New York: McGraw-Hill, 1960, chap. 1.

CHAPTER 8 GRAPHICAL METHODS

8–1. *THE USE OF GRAPHICAL METHODS*

We know, in principle, how to solve any equilibrium problem. We simply write the required number of equilibrium constant expressions, material balance conditions, and the ionic balance expression, manipulate them algebraically to obtain one equation with one unknown, and solve this equation. Life is really not that simple, however. It is often quite difficult to combine all of the equations into a single expression. Even when we have one equation with one unknown, that equation may be difficult to solve (e.g., a fifth or sixth degree equation). Sometimes we can guess reasonable approximations to make the mathematics easier. Often, especially in complicated problems, we cannot guess which species are negligible and which are not. Under these circumstances graphical methods are valuable. In a graphical solution of a problem each equilibrium constant expression is represented by a single curve on a graph. After all the curves are plotted, and on the basis of the other information in the problem, we can usually approximate an answer and tell at a glance which species are present in negligible amounts and which species are not. If the reader is not convinced of the utility of this kind of approach, let him try to calculate the pH of a solution of 0.10 mole of $(NH_4)_2HPO_4$ diluted to 1 liter with water (knowing K_1, K_2, and K_3 for H_3PO_4 and K_a for NH_4^+). Graphical methods are also applicable in a rapid survey of a system. Once we draw a graph of a system we can predict how it will behave under a variety of conditions.

There are several methods that can be used to solve problems graphically. The most useful one is the *master variable* technique which has been applied quite extensively in the Scandinavian countries, but is not often mentioned in the American literature.*. A straight line is the most convenient curve to plot since only two points define the shape of the whole curve. The student may remember that the equation of a straight line has the form

$$x + y = c \qquad (8.1)$$

Equilibrium constant expressions on the other hand are of the form

$$xy = z \qquad (8.2)$$

that is, in the form of products rather than sums. To put an equililibrium constant expression into a linear form we can take logarithms and convert it to the form

$$\log x + \log y = \log z \qquad (8.3)$$

If all of this seems a bit obscure at the moment, it is only because we are talking in generalities. The method will become clearer when we get to some specific examples. To sum up before we use the method, we will choose a *master variable*, that is the concentration of the species which seems to be the important one, e.g., $[H^+]$ in acid-base problems. We then draw curves to show how the other concentrations vary as we vary the master variable. In order to linearize our curves we plot logarithms of the concentrations.

8–2. ACID-BASE PROBLEMS

LOG C vs. pH

Let us use a graphical method to solve some problems concerning a simple system first; a 0.10 M solution of an acid, HA, $K_a = 1.0 \times 10^{-5}$.

---

* An excellent review of graphical methods is that of K. G. Sillen in I. M. Kolthoff and P. J. Elving, *Treatise on Analytical Chemistry*, New York: Interscience, 1959, Part I, Vol. 1, pp. 277–317.

The species that are present in the solution are H^+, OH^-, HA, and A^-. We want to plot the log of the concentrations of each of these species against the log of the concentration of the master variable, which in acid-base problems is almost always log $[H^+]$ or pH. Therefore we are going to plot log $[H^+]$, log $[OH^-]$, log $[HA]$, and log $[A^-]$ against pH (Fig. 8.1). A plot of log $[H^+]$ vs. pH is easy, since

$$\log [H^+] = -pH \tag{8.4}$$

When the pH is 3, log $[H^+] = -3$, when the pH is 9, log $[H^+] = -9$ etc. (8.4) is then a straight line, called line 1 in Figure 8.1. To plot log

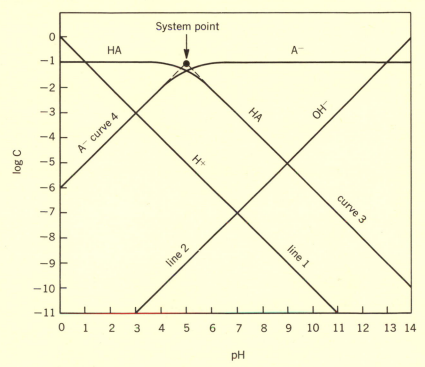

FIGURE 8.1. Master variable diagram for the system HA— A^-. $K_a = 1.0 \times 10^{-5}$, $C_A = 0.1\ M$.

$[OH^-]$ vs. pH we use the K_w expression

$$[H^+][OH^-] = K_w = 10^{-14} \tag{8.5}$$

or

$$\log [OH^-] = -14 - \log [H^+] = -14 + pH \tag{8.6}$$

When the pH is 4.0, log $[OH^-] = -10$, when the pH is 11, log $[OH^-] = -3$, etc. (8.6) also represents a straight line, called line 2 in Figure 8.1.

To plot log [HA] and log [A⁻] we have to use the expression

$$\frac{[H^+][A^-]}{[HA]} = K_a = 10^{-5} \tag{8.7}$$

and also the material balance equation

$$[HA] + [A^-] = C_A = 0.10 \tag{8.8}$$

We can solve (8.7) and (8.8) for [HA] as a function of $[H^+]$ and for [A⁻] as a function of $[H^+]$. From (8.7)

$$[A^-] = \frac{K_a[HA]}{[H^+]} \tag{8.9}$$

Introducing (8.9) into (8.8)

$$[HA]\left[1 + \frac{K_a}{[H^+]}\right] = C_A \tag{8.10}$$

or
$$[HA] = \frac{[H^+]C_A}{[H^+] + K_a} \tag{8.11}$$

Taking logarithms of (8.11) we have an expression for log [HA] in terms of log $\{[H^+] + K_a\}$, pH, and log C_A. However (8.11) represents a curve rather than a straight line. Let us plot some points along the curve. When the pH is 2, $[H^+] = 10^{-2}$, and

$$[HA] = \frac{(10^{-2})(10^{-1})}{10^{-2} + 10^{-5}} \approx 10^{-1}$$

$$\log [HA] = -1$$

This will also be true at all pH's more acidic than this, e.g., pH = 1, as the reader can verify. This implies that in very acidic solutions essentially all the acid is in the form of HA. When the pH is 9, $[H^+] = 10^{-9}$, and

$$[HA] = \frac{(10^{-9})(10^{-1})}{10^{-9} + 10^{-5}} \approx 10^{-5}$$

$$\log [HA] = -5$$

At a pH of 10, by a similar calculation, we find that log [HA] = −6. In alkaline solution, a ten-fold decrease in the $[H^+]$ (a pH increase of 1) brings about a ten-fold decrease in the [HA]. At pH = 5, $[H^+] = 10^{-5}$,

$$[HA] = \frac{(10^{-5})(10^{-1})}{10^{-5} + 10^{-5}} = \frac{10^{-1}}{2} = 5 \times 10^{-2}$$

$$\log [HA] = -1.3$$

A point-by-point plot of (8.11) leads to curve 3 in Figure 8.1. In a similar way we can solve (8.7) for $[A^-]$

$$[HA] = \frac{[H^+][A^-]}{K_a} \tag{8.12}$$

and introducing (8.12) into (8.8)

$$[A^-]\left[1 + \frac{[H^+]}{K_a}\right] = C_A \tag{8.13}$$

$$[A^-] = \frac{K_a C_A}{[H^+] + K_a} \tag{8.14}$$

We can now calculate a curve for log $[A^-]$ vs. pH, just as we did before, point-by-point. By this method, curve 4 in Figure 8.1 is obtained.

Some Simplifications

Now that we see how the curves look, we can make our calculations easier by the following approach. Let us see how the equation

$$[HA] = \frac{[H^+]C_A}{[H^+] + K_a} \tag{8.12}$$

behaves at the two extremes of $[H^+]$. When the solution is acidic, the $[H^+]$ is large, and when $[H^+] \gg K_a$, so that $[H^+] + K_a \approx [H^+]$, (8.12) simply becomes

$$[HA] = C_A \quad \text{or} \quad \log [HA] = \log C_A$$

On the other hand, when the $[H^+]$ is small, and $[H^+] \ll K_a$, then $[H^+] + K_a \approx K_a$, and (8.12) becomes

$$[HA] = \frac{[H^+]C_A}{K_a}$$

$$\log [HA] = \log C_A - \log K_a - pH$$

or in this case

$$\log [HA] = -1 - (-5) - pH = 4 - pH$$

This is the straight line portion of curve 3 at high pH's. Only at intermediate values of pH, when the $[H^+]$ is of the order of K_a, i.e., $[H^+]$ between 10^{-4} and 10^{-6} M, does the curve deviate significantly from linearity. In this region we have a simple point to remember, however. When the $[H^+] = K_a$, half of the acid is in the form of HA and half is in the form of A^-, so that

$$[HA] = \tfrac{1}{2}C_A \quad \text{or} \quad \log [HA] = \log C_A - 0.3$$

or in this case
$$\log [HA] = -1 - 0.3 = -1.3$$

With a little practice we can sketch the curve for log [HA] fairly rapidly: it is constant and equal to log C_A at low pH's, linear and decreasing one log C unit per one pH unit increase, and passing through the *system point* (the point $\log C = \log C_A$, $pH = -\log K_a$), and finally, equal to $\log C_A - 0.3$ at pH $= -\log K_a$ (or 0.3 units directly below the system point). By exactly analogous reasoning, from (8.14), we find that when the [H$^+$] is small

$$\log [A^-] = \log C_A$$

and when the [H$^+$] is large

$$\log [A^-] = \log C_A + \log K_a + pH$$

or in this case

$$\log [A^-] = -1 + (-5) + pH = -6 + pH$$

Summary

Let us summarize the procedure for drawing a master variable graph for an acid-base system.

(1) Draw axes with pH as the abscissa (the horizontal or X axis) and log C as the ordinate (the vertical or Y axis).

(2) Draw the straight lines representing $\log [H^+]$ vs. pH and log [OH$^-$] vs. pH.

(3) Add the *system point*, $pH = -\log K_a$, $\log C = \log C_A$.

(4) Draw the curve for HA. At acidic pH's, $\log [HA] = \log C_A$, and at basic pH's, log [HA] decreases one unit per unit increase in pH (i.e., has a slope of -1), and would intersect the system point if extended. At intermediate pH's, the curve passes through a point 0.3 units below the system point.

(5) Draw the curve for A$^-$. At acidic pH's, log [A$^-$] increases one unit per unit increase in pH (i.e., has a slope of $+1$) and would intersect the system point if extended. At basic pH's, $\log [A^-] = \log C_A$. At intermediate pH's, the curve passes through the point 0.3 units below the system point.

APPLICATIONS OF THE ACID-BASE GRAPH

Now that we have our graph we can use it to solve any problem concerning solutions containing HA and A$^-$ at a total concentration of 0.1 *M*.

Example 8.1. Calculate the [HA] and the [A⁻] at pH = 2.5.

By inspection of the graph, we see that at pH = 2.5

$$\log [HA] = -1.0 \quad \text{or} \quad [HA] = 10^{-1} \, M$$

$$\log [A^-] = -3.5 = -4.0 + 0.5$$

$$[A^-] = 3.2 \times 10^{-4} \, M \quad \textit{Answer}$$

We can calculate the concentration of HA and A⁻ at any other pH by just this technique. Before we attempt to solve any other problems let us introduce the concept of the *proton condition*. Assume we add pure HA to H_2O. We will call HA and H_2O the *zero level*. When HA and H_2O dissociate to form A⁻, H⁺, and OH⁻, the sum total of the concentrations of all species containing more protons than species at the zero level must equal the sum total of the concentrations of species containing less protons than the zero level. In this example, only one substance contains more protons than the zero level, H⁺ (i.e., H_3O^+) while A⁻ and OH⁻ contain less protons than the zero level. Therefore

$$[H^+] = [A^-] + [OH^-] \qquad (8.15)$$

This equation is the proton condition for the system HA + H_2O.

Example 8.2. Calculate the [H⁺], [OH⁻], [HA], and [A⁻] in a solution prepared by diluting 0.1 mole of HA to 1 liter with water.

This is the familiar Type 1 problem. The proton condition for this problem is given in (8.15), or

$$\log [H^+] = \log ([A^-] + [OH^-]) \qquad (8.16)$$

The solution to the problem is the location on the graph at which (8.16) is satisfied. Looking along the H⁺ line (line 1) in Figure 8.1, we see that at almost all points along this line the [A⁻] ≫ [OH⁻], or line 4 lies above line 2. Therefore we can neglect the [OH⁻] and find the point where

$$\log [H^+] = \log [A^-]$$

or the intersection of lines 1 and 4. By inspection then:

$$\log [A^-] = -3 \qquad [A^-] = 10^{-3} \, M$$
$$pH = 3.0 \qquad [H^+] = 10^{-3} \, M$$
$$\log [OH^-] = -11 \qquad [OH^-] = 10^{-11} \, M$$
$$\log [HA] = -1 \qquad [HA] = 10^{-1} \, M$$

Example 8.3. Calculate the [H⁺], [OH⁻], [HA], and [A⁻] in a solution prepared by diluting 0.1 mole of NaA to 1 liter with water.

Our zero level in this problem is A^- and H_2O. Species containing more protons than the zero level are HA and H^+. The species containing less protons is OH^-. Therefore the proton condition is

$$[H^+] + [HA] = [OH^-] \qquad (8.17)$$

Again, by inspection of the graph, we see that the HA line is above the H^+ line, which means that the $[HA] \gg [H^+]$, and we look for the point where

$$\log [HA] = \log [OH^-]$$

the intersection of lines 2 and 3. By inspection

$$\log [HA] = -5 \qquad [HA] = 10^{-5} \, M$$
$$pH = 9.0 \qquad [H^+] = 10^{-9} \, M$$
$$\log [OH^-] = -5 \qquad [OH^-] = 10^{-5} \, M$$
$$\log [A^-] = -1 \qquad [A^-] = 10^{-1} \, M$$

The reader might question the procedure of neglecting the lower lines. Suppose we wanted to find the point where the complete condition in Example 8.3

$$\log ([H^+] + [HA]) = \log [OH^-]$$

is met exactly. The H^+ line lies vertically 4 log C units below the HA line, so that

$$[H^+] = 10^{-4} [HA]$$
$$[H^+] + [HA] = [HA](1 + 10^{-4}) = 1.0001 [HA]$$

Obviously, in this case, the contribution of H^+ to the sum $[H^+] + [HA]$ is negligible. Suppose the lines were closer together. Suppose, for example, that the H^+ line lay only 0.2 of a unit below the HA line. Then

$$[H^+] = 10^{-0.2} [HA] = 0.625 [HA]$$
$$[H^+] + [HA] = 1.625 [HA] = 10^{0.21} [HA]$$
$$\log [OH^-] = \log ([H^+] + [HA]) = 0.21 + \log [HA]$$

so that the correct intersection point would be 0.21 of a log C unit above the point chosen and the correct answer would be pH $= 9.10$ rather than pH $= 9.00$. Table 8.1 shows the number of units that the actual intersection line will lie above the upper line of a pair, when the upper line is various distances above the lower line. For all practical calculations, we can, as a first approximation, always neglect the lower lines and the maximum error made by this procedure will be about 0.15 pH or log C unit.

TABLE 8.1. Correction Table for Contributions
of Lower Lines

If the lower line lies this many units below the upper line ...	then the new intersection line should be constructed this many units above upper line
0.0	0.30
0.1	0.26
0.2	0.21
0.3	0.18
0.4	0.15
0.5	0.12
0.6	0.10
0.7	0.08
0.8	0.06
0.9	0.05
1.0	0.04

MIXTURES OF ACIDS AND BASES

If more than one acid-base system is present in a solution, we can add lines for the additional systems to our graph and solve problems by the techniques shown above.

Example 8.4.   Calculate the pH of a 0.1 M NH$_4$A solution.

This problem involves the additional acid-base system, NH_4^+—NH_3. First we add the lines corresponding to the new system to those already present for the HA—A$^-$ system (Figure 8.2). These lines are plotted by exactly the same technique as before. We add the system point, pH $= 9.3$, log $C = -1$ (since K_a of $NH_4^+ = 5 \times 10^{-10}$ and $C_{NH_4} = 0.1$ M) and draw curves 5 and 6. Now we define our proton condition. Since NH$_4$A was added to water, the zero level is NH_4^+, A$^-$, and H$_2$O. Species that have an excess of protons above the zero level are H$^+$ and HA. Species having a deficiency of protons below the zero level are OH$^-$ and NH$_3$. The proton condition is

$$[H^+] + [HA] = [OH^-] + [NH_3]$$

From the graph we see that the HA curve lies well above the H$^+$ line and the NH$_3$ curve lies well above the OH$^-$ line, so that

$$[HA] \gg [H^+] \quad \text{and} \quad [NH_3] \gg [OH^-]$$

and the proton condition becomes

$$[HA] = [NH_3] \quad \text{or} \quad \log[HA] = \log[NH_3]$$

The desired point on the graph is the intersection of lines 3 and 6, and we find

$$pH = 7.15 \quad \textit{Answer}$$

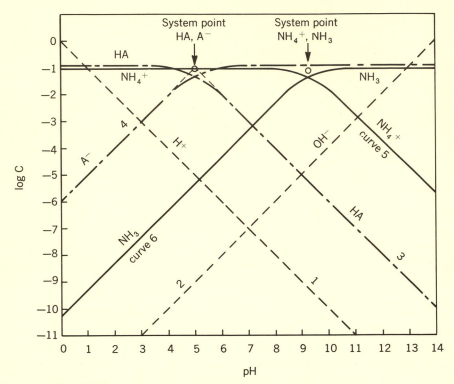

FIGURE 8.2. Master variable diagram for the systems HA—A$^-$: $K_a = 1.0 \times 10^{-5}$, $C_A = 0.1$ M and NH$_4^+$—NH$_3$: $K_a = 5 \times 10^{-10}$, $C_{NH_3} = 0.1$ M.

POLYPROTIC ACIDS

Polyprotic acids are essentially equivalent to a mixture of acids and their graphs are plotted in a similar manner. For example, consider a 0.1 M solution of the acid H$_2$X, $K_1 = 10^{-3}$ and $K_2 = 10^{-7}$. Without going through a mathematical analysis of the system, which is similar to that of monoprotic acids, we can state that there are two system points: pH = 3, log $C = -1$ (point a) and pH = 7, log $C = -1$ (point b). In very acidic solutions, H_2X predominates, and is present at a concentration of essentially 0.1 *M* (curve 1, Fig. 8.3). To the right of system point a, log $[H_2X]$ decreases with a slope of -1. HX^- predominates only in the zone between the system points and decreases to the left of point a and to the right of point b (curve 2). X^{--} predominates at basic pH's (to the right of point b) and decreases to the left of point b

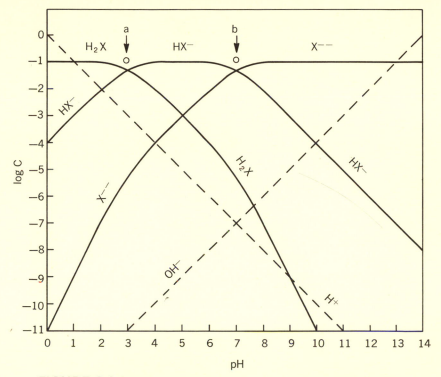

FIGURE 8.3. Master variable diagram for the system H_2X—HX^-
—X^{--}: $K_1 = 1.0 \times 10^{-3}$, $K_2 = 1.0 \times 10^{-7}$, $C_A = 0.1$ *M*.

with a slope of $+1$. Actually, when the X^{--} curve extends to the pH region where H_2X predominates, its slope changes to $+2$. Similarly, when the H_2X curve reaches the region where X^{--} predominates, its slope changes to -2. These changes are shown on the graph, but are of little practical importance, since the intersections of interest usually lie well above these points. Figure 8.3, with the usual lines for log $[H^+]$ and log $[OH^-]$, represents a graphical representation of the H_2X—HX^-—X^{--} system.

Example 8.5. Calculate the pH of a 0.1 *M* H_2X solution. What is the $[X^{--}]$ in this solution?

The zero level is H_2X and H_2O, and the proton condition is

$$[H^+] = [HX^-] + 2[X^{--}] + [OH^-]$$

Note that a 2 appears in front of the $[X^{--}]$ term, because this species contains two protons less than the zero level. Similarly, a species containing two protons above the zero level would be multiplied by 2. Looking at intersections with the H^+ line, we see that the X^{--} and OH^- curves lie far below the HX^- curve, so that

$$[HX^-] \gg [X^{--}] + [OH^-]$$

and $[HX^-] + 2[X^{--}] + [OH^-] \approx [HX^-]$

The proton condition becomes

$$\log [H^+] = \log [HX^-]$$

which occurs at a pH $= 2.0$. *Answer*

At this pH

$$\log [X^{--}] = -7.0$$

$$[X^{--}] = 1.0 \times 10^{-7}\, M \quad Answer$$

Notice that the approximations assumed in the numerical solution of diprotic acid problems can be seen quite readily in the graph of the system.

Example 8.6. Calculate the pH of a 0.1 M NaHX solution.

The zero level is HX^- and H_2O, and the proton condition is

$$[H^+] + [H_2X] = [X^{--}] + [OH^-]$$

From the graph we see that the $[H_2X] \gg [H^+]$, and the $[X^{--}] \gg [OH^-]$. The intersection of interest is

$$\log [H_2X] = \log [X^{--}]$$

$$pH = 5.0 \quad Answer$$

Example 8.7. Calculate the pH of a 0.1 M Na$_2$X solution.

The zero level is X^{--} and H_2O and the proton condition is

$$[H^+] + [HX^-] + 2[H_2X] = [OH^-]$$

Since the $[HX^-] \gg [H_2X] + [H^+]$, the required intersection is

$$\log [HX^-] = \log [OH^-]$$

$$pH = 10.0 \quad Answer$$

Example 8.8. Calculate the pH of a solution prepared by diluting 0.1 mole of $(NH_4)_2X$ to 1 liter with water.

Since we have added another acid-base system $(NH_4^+—NH_3)$ to the problem, we must draw the curves to represent it on our graph (Figure 8.4). Note that the system point for the $NH_4^+—NH_3$ system is

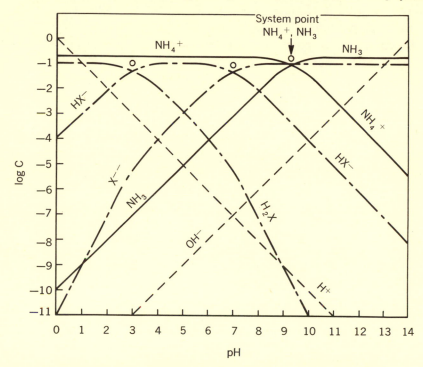

FIGURE 8.4. Master variable diagram for the systems $H_2X—$ $HX^-—X^{--}$: $K_1 = 1.0 \times 10^{-3}$, $K_2 = 1.0 \times 10^{-7}$, $C_A = 0.1$ M. $NH_4^+—NH_3$: $K_a = 5 \times 10^{-10}$, $C_{NH_3} = 0.2$ M.

pH $= 9.3$, log C $= -0.7$, because the total concentration of ammonia species is 0.2 M (and log 0.2 $= -0.7$). The zero level condition is

$$[H^+] + [HX^-] + 2[H_2X] = [NH_3] + [OH^-]$$

With the approximations

$$[HX^-] \gg [H_2X] + [H^+] \quad \text{and} \quad [NH_3] \gg [OH^-]$$

The required intersection occurs where

$$\log [HX^-] = \log [NH_3]$$

$$pH = 8.0 \quad \textit{Answer}$$

The reader should appreciate that this last problem was by no means a simple one and the apparent ease of solution demonstrates the utility of the graphical approach to solving problems. When the system points are near one another, so that lines lie close together, more careful plotting of the different lines (by point-by-point calculations from the equilibrium constant expressions) may be necessary. But, once the graph has been plotted, calculations are made as before and close-lying line corrections, when necessary, can be made using the factors given in Table 8.1. These methods can also be extended to problems involving triprotic acids, mixtures of acids, etc., by exactly the same techniques.

DISTRIBUTION DIAGRAMS

Another diagram frequently used to represent acid-base systems and complex ion systems, is the *distribution* diagram. This diagram shows the fraction of the total amount of a substance in a given form vs. some variable, such as pH. For example, Figure 8.5 shows the

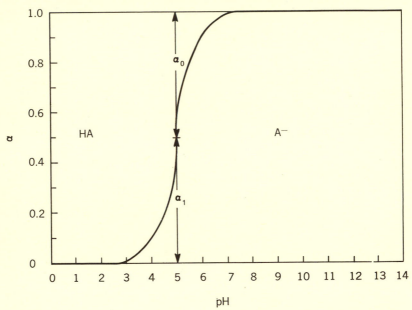

FIGURE 8.5. Distribution diagram for the system HA—A$^-$, $K_a = 1.0 \times 10^{-5}$. α_0 is the fraction of the total acid present as HA, and α_1 is fraction present as A$^-$.

distribution diagram for the acid HA. A distribution diagram can easily be calculated from the master variable diagram by noting the concentration of each species at a given pH and then calculating the fraction of each species. For example, at a pH of 4, referring to Figure 8.1, we find that the $[HA] = 0.09\ M$ and the $[A^-] = 0.01\ M$, so that the fraction of the total amount of acid present as the singly ionized form A^-, α_1, is

$$\alpha_1 = \frac{[A^-]}{[A^-] + [HA]} = \frac{0.01}{0.10} = 0.1$$

Point-by-point plotting, at various pH's, yields the distribution diagram. Figure 8.6 is the distribution diagram for H_2X.* Distribution diagrams

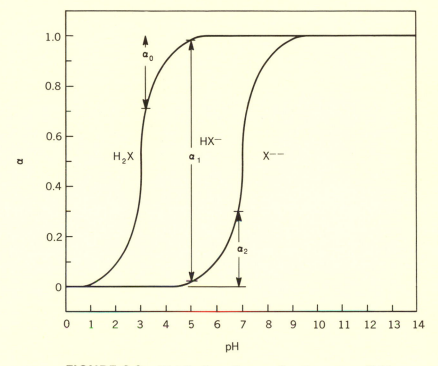

FIGURE 8.6. Distribution diagram for the system H_2X—HX^-—X^{--}, $K_1 = 1.0 \times 10^{-3}$, $K_2 = 1.0 \times 10^{-7}$. α_0 is the fraction of the total acid present as H_2X, α_1 is the fraction present as HX^-, and α_2 is the fraction present as X^{--}.

* An alternate method of plotting distribution diagrams involves plotting a line representing the fraction of each species, so that there would be a line for H_2X, a line for HX^- (which would increase, go through a peak at $pH = 5$, and then decrease again), and a line for X^{--}. The method we have chosen, plotting the lines as boundaries between zones so that the fraction of the species present as HX^-, α_1, is the distance between the two boundary lines, has the advantage of using fewer lines, and in situations where there are many species, giving a less-complicated graph.

are convenient for visualizing a system, and show at a glance which species predominate. They cannot, however, be used to solve most problems and they actually provide less information than master variable diagrams do.

8-3. SOLUBILITY PRODUCT PROBLEMS

MASTER VARIABLE DIAGRAMS

Graphical methods can also be applied to solubility equilibria. For example, suppose we want to graph the system $Mg(OH)_2$—Mg^{++}— OH^- (K_{sp} of $Mg(OH)_2$ is 1×10^{-11}). We can choose as our master variable pH (or, equally well, pOH). The system is governed by the solubility product expression:

$$[Mg^{++}][OH^-]^2 = 1 \times 10^{-11} \qquad (8.18)$$

We now calculate $\log [Mg^{++}]$ and $\log [OH^-]$ as a function of pH. Using the K_w expression, as before, we calculate $\log [OH^-]$ as a function of pH as

$$\log [OH^-] = -14 + pH \qquad (8.6)$$

To calculate $\log [Mg^{++}]$ we take the logarithm of (8.18)

$$\log [Mg^{++}] + 2 \log [OH^-] = -11.0$$

$$\log [Mg^{++}] = -11.0 - 2 \log [OH^-] = -11.0 + 28.0 - 2\,pH$$

$$\log [Mg^{++}] = 17.0 - 2\,pH \qquad (8.19)$$

We can plot this line by calculating any two points that satisfy (8.19) and joining them. For example, at $pH = 9$

$$\log [Mg^{++}] = 17 - 2(9) = -1 \quad \text{etc.}$$

The master variable diagram for the $Mg(OH)_2$—Mg^{++}—OH^- system is shown in Figure 8.7.

Example 8.9. A solution initially contains 0.1 M Mg^{++}. What concentration of Mg^{++} remains unprecipitated when the solution is brought to a pH of 10? What fraction of the Mg^{++} precipitated?

From Figure 8.7 we see that at $pH = 10$, $\log [Mg^{++}] = -3.0$, or

$$[Mg^{++}] = 1.0 \times 10^{-3} M \quad Answer$$

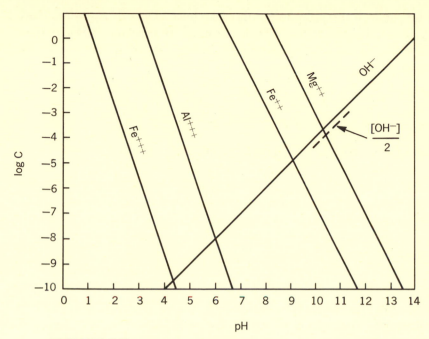

FIGURE 8.7. Master variable diagram for several metal hydroxides.

The fraction of the Mg^{++} unprecipitated is then

$$\frac{1.0 \times 10^{-3}}{0.1} = 0.01$$

so that the fraction precipitated is $1.00 - 0.01 = 0.99$.

Example 8.10. When an excess of solid $Mg(OH)_2$ is equilibrated with water, what will the concentrations of Mg^{++} and OH^- be in the resulting solution?

This is a Type 1 solubility problem. To solve it, we must write some equation involving Mg^{++} and OH^- satisfying the conditions of the problem (just as we wrote the proton condition in acid-base problems). For example, the electroneutrality condition in this solution is

$$[Mg^{++}] = \tfrac{1}{2}[OH^-]$$

or

$$\log[Mg^{++}] = \log[OH^-] - \log 2$$
$$\log[Mg^{++}] = \log[OH^-] - 0.3$$

We must find on the graph the point which satisfies this equation. The broken line lying slightly below the OH^- line is the line, $\log[OH^-] - 0.3$, and it therefore crosses the $\log[Mg^{++}]$ line at the solution of the problem. From the graph we find

$$[Mg^{++}] = 1.3 \times 10^{-4} \, M \qquad [OH^-] = 2.6 \times 10^{-4} \, M \quad Answer$$

SEPARATIONS

We have plotted, in Figure 8.7, lines representing the solubility product constants of $Fe(OH)_2$, $Fe(OH)_3$, and $Al(OH)_3$. The reader may confirm that these lines have been plotted correctly. Using diagrams such as these, involving several systems, we can predict the success of separations of metal ions by precipitation of one and not the other, and we can calculate the conditions under which such separations will be complete.

Example 8.11. An analyst has a solution which contains Al^{+++} and Fe^{+++}, each at a concentration of $10^{-2} \, M$. At what pH will the separation of these metals, by the precipitation of only $Fe(OH)_3$, be most complete? What concentration of Fe^{+++} will remain unprecipitated at this pH?

Looking at Figure 8.7, we see that as long as the pH is smaller than 1.5 neither metal ion will precipitate. At a pH of about 1.8, $Fe(OH)_3$ starts to precipitate from a $10^{-2} \, M$ solution ($\log[Fe^{+++}] = -2$). Al^{+++} doesn't start to precipitate from this solution until the pH is about 4. If the pH is fixed at about 4, no Al^{+++} will precipitate, and the $[Fe^{+++}]$ will be decreased to (reading along the pH = 4 line) 2.5×10^{-9} M (i.e., $\log[Fe^{+++}] = -8.6$). Therefore a successful separation of Fe^{+++} from Al^{+++} in this solution can best be accomplished by adjusting the pH to about 4 (or slightly more acidic to be on the safe side).*

Graphical methods can also be applied to other precipitates, such as CaF_2. The K_{sp} expression for CaF_2 is

$$[Ca^{++}][F^-]^2 = K_{sp} = 4 \times 10^{-11} \tag{8.20}$$

To draw a graph of this system we must first choose a master variable. The obvious choice in this case is either pCa^{++} or pF^- (remembering

---

* The reader should realize that other factors may come into play which make a separation less successful than that predicted by the K_{sp} calculations. Aside from activity corrections, which might cause our calculations to be slightly in error, factors like entrapment of some Al^{+++} inside or on the surface of the precipitate (called *coprecipitation*) may cause the separation to fail. Nevertheless, the calculation just performed would give the experimenter a basis for starting some experiments to see if the separation were possible.

that "p" is an abbreviation of $-\log$). Suppose we take pF^- as the master variable. Taking the logarithm of (8.20)

$$\log [Ca^{++}] + 2 \log [F^-] = -10.4$$

$$\log [Ca^{++}] = -10.4 - 2 \log [F^-] = -10.4 + 2 pF^- \qquad (8.21)$$

The $\log C$ vs. pF^- diagram for this system is shown in Figure 8.8.

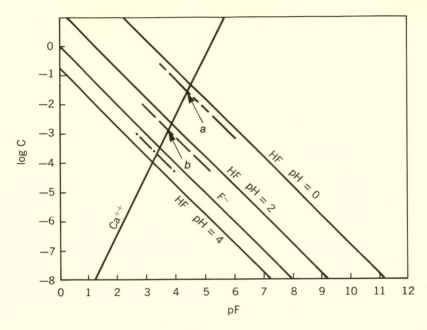

FIGURE 8.8. Master variable diagram for the system $CaF_{2(solid)}$ —Ca^{++}—F^-—HF.

Example 8.12. Calculate the $[Ca^{++}]$ in a solution prepared by shaking an excess of solid CaF_2 with water (neglecting the bacisity of F^-).
Since only CaF_2 dissolves,

$$CaF_{2(solid)} \rightleftarrows Ca^{++} + 2F^-$$

and, by the electroneutrality condition,

$$[Ca^{++}] = \tfrac{1}{2}[F^-]$$

$$\log [Ca^{++}] = \log [F^-] - 0.3$$

$$\log [Ca^{++}] = -3.7 = -4.0 + 0.3$$

$$[Ca^{++}] = 2 \times 10^{-4} M \quad \textit{Answer}$$

Problems involving an excess of F^- (Type 2 problems) can be solved by calculating the prevailing pF^- and reading the corresponding $[Ca^{++}]$ on the graph.

8–4. COMPETING SOLUBILITY AND ACID-BASE EQUILIBRIA

Master variable diagrams can be used to solve problems involving the solubility of a precipitate containing an anion base, such as F^-, in acidic solutions. Additional lines are added to the graph of the system to represent any additional equilibria.

Example 8.13. Calculate the solubility of CaF_2 in a solution of $pH = 0$ (e.g., 1 M HCl).

The solubility of CaF_2 will be greater in this acidic solution because of the reaction of F^-, formed when CaF_2 dissolves, with H^+ to form weakly ionized HF. We must consider then the K_a expression

$$\frac{[H^+][F^-]}{[HF]} = K_a = 6 \times 10^{-4} \tag{8.22}$$

and add a line to our diagram for log [HF]. Taking the logarithm of (8.22) we obtain

$$\log [H^+] + \log [F^-] - \log [HF] = -3.22$$

or

$$\log [HF] = 3.22 - pF^- - pH \tag{8.23}$$

In this case, since the $pH = 0$, the expression becomes

$$\log [HF] = 3.22 - pF^-$$

This line is shown in Figure 8.8.

To solve the problem, we make use of the material balance equation. Since all the Ca^{++} and all the different forms of fluoride come from the CaF_2 (with two fluorides being produced per calcium)

$$[Ca^{++}] = \tfrac{1}{2}([F^-] + [HF])$$

or

$$\log [Ca^{++}] = \log ([F^-] + [HF]) - 0.3 \tag{8.24}$$

Since the HF line lies well above the F^- line at a pH of 0,

$$[HF] \gg [F^-] \quad \text{or} \quad [HF] + [F^-] \approx [HF]$$

and we look for the intersection where

$$\log [Ca^{++}] = \log [HF] - 0.3 \qquad (8.25)$$

This intersection is marked point a in Figure 8.8, and

$$\log [Ca^{++}] = -1.55$$

$$\text{solubility} = [Ca^{++}] = 3 \times 10^{-2} M \quad Answer$$

Example 8.14. Calculate the solubility of CaF_2 in a solution of pH $= 2$ (e.g., 0.01 M HCl).

Application of (8.23) indicates that the log [HF] line is given by the equation

$$\log [HF] = 1.22 - pF^-$$

This line is also plotted in Figure 8.8. Again we can apply (8.24) and solve the problem by neglecting the $[F^-]$.* The result is the intersection marked b, and

$$\log [Ca^{++}] = -2.9$$

$$\text{solubility} = [Ca^{++}] = 1.3 \times 10^{-3} M \quad Answer$$

Example 8.15. Calculate the solubility of CaF_2 in a solution of pH $=4$.

Using the procedure outlined above, we can construct the line for log [HF] at a pH of 4. In this case,

$$[F^-] > [HF] \quad \text{and} \quad \log ([HF] + [F^-]) \approx \log [F^-]$$

so that the result in this case is essentially the same as that in Example 8.12.

In conclusion, the procedure for solving these problems involves (1) devising a method of plotting a line representing the desired equilibrium vs. the master variable, and (2) formulating some expression containing the concentrations of the species in the solution that may be applied to the diagram. Of course there are some problems in which the required relations are not as obvious as in the preceding examples. Suppose, for example, we want to calculate the solubility of CaF_2 in water, not neglecting the basicity of F^-. In this situation neither the [HF] nor the $[H^+]$ (nor $[OH^-]$) is known. How would we proceed? One approach would be to write the additional equations of interest

$$\frac{[OH^-][HF]}{[F^-]} = K_b = \frac{K_w}{K_a} \qquad (8.26)$$

* The F^- line, about 1.2 units below the HF line, will contribute less than 0.04 units to the HF line (see Table 8.1), and is, for all practical purposes, negligible.

and the ionic and material balance equations (neglecting H^+)

$$2[Ca^{++}] = [F^-] + [OH^-] \tag{8.27}$$

$$2[Ca^{++}] = [F^-] + [HF] \tag{8.28}$$

Combining (8.27) and (8.28) we obtain

$$[HF] = [OH^-] \tag{8.29}$$

and, using (8.26)

$$\frac{[OH^-]^2}{[F^-]} = K_b = \frac{K_w}{K_a} \tag{8.30}$$

The logarithm of (8.30) provides the needed expression for plotting log $[OH^-]$ vs. pF^-, and (8.27) provides the expression for locating the desired intersection point. The completion of this problem and the proof that the basicity of the fluoride ion is not significant in this case is left to the reader.

8–5. COMPLEX ION PROBLEMS

MASTER VARIABLE DIAGRAMS

In our previous treatment of complex ions (Chapter 5) we discussed only those cases in which the complexing agent was present in such a large excess that only the most highly coordinated species of the complex ion had to be considered. For example, we considered the case of Cu^{++} in the presence of a large excess of ammonia, so that the predominant species was $Cu(NH_3)_4^{++}$, and we only had to consider the equilibrium

$$Cu^{++} + 4NH_3 \rightleftarrows Cu(NH_3)_4^{++} \tag{8.31}$$

We may ask, under what conditions is $Cu(NH_3)_4^{++}$ the only significant form of copper-ammonia complex present, and under what conditions will the other, lower species, $Cu(NH_3)_3^{++}$, $Cu(NH_3)_2^{++}$, and $Cu(NH_3)^{++}$, be present in large amounts? To answer these questions we must consider the additional equilibria

$$Cu^{++} + 3NH_3 \rightleftarrows Cu(NH_3)_3^{++} \tag{8.32}$$

$$Cu^{++} + 2NH_3 \rightleftarrows Cu(NH_3)_2^{++} \tag{8.33}$$

$$Cu^{++} + NH_3 \rightleftarrows Cu(NH_3)^{++} \tag{8.34}$$

each represented by an equilibrium constant expression.* Our problem then is to solve the following equilibrium constant expressions for the concentration of each species as a function of the NH_3 concentrations.

$$\frac{[Cu(NH_3)^{++}]}{[Cu^{++}][NH_3]} = 1.35 \times 10^4 \tag{8.35}$$

$$\frac{[Cu(NH_3)_2^{++}]}{[Cu^{++}][NH_3]^2} = 4.1 \times 10^7 \tag{8.36}$$

$$\frac{[Cu(NH_3)_3^{++}]}{[Cu^{++}][NH_3]^3} = 3.0 \times 10^{10} \tag{8.37}$$

$$\frac{[Cu(NH_3)_4^{++}]}{[Cu^{++}][NH_3]^4} = 3.9 \times 10^{12} \tag{8.38}$$

Although these calculations can be performed without too much mathematical difficulty, a graphical approach is easier. We choose as our master variable $pNH_3(-\log[NH_3])$, and now calculate the concentration of each species. One approach would be to treat this problem as a polyprotic acid problem (writing stepwise formation constants) and drawing curves similar to those of a polyprotic acid. Since the formation constants are usually numerically quite close together, this graph can be messy and tedious to construct. Let us try another approach. Let us arbitrarily assume a concentration of uncomplexed Cu^{++} of, for example, 10^{-8} M. Once this concentration is fixed, all of the above equations become straight lines on a $\log C$—pNH_3 diagram. Taking the logarithm of (8.35) we obtain

$$\log[Cu(NH_3)^{++}] = 4.13 + \log[Cu^{++}] + \log[NH_3]$$

or, since $\log[Cu^{++}] = -8$ and $pNH_3 = -\log[NH_3]$,

$$\log[Cu(NH_3)^{++}] = -3.87 - pNH_3 \tag{8.39}$$

This equation is plotted as line 1 in Figure 8.9. Similarly, from the other equations we obtain

$$\log[Cu(NH_3)_2^{++}] = -0.39 - 2pNH_3 \tag{8.40}$$

$$\log[Cu(NH_3)_3^{++}] = 2.48 - 3pNH_3 \tag{8.41}$$

$$\log[Cu(NH_3)_4^{++}] = 4.59 - 4pNH_3 \tag{8.42}$$

and we can plot lines 2, 3, and 4. We can tell at a glance from this graph

---

* Rather than considering the *overall* formation equilibria, we could alternately consider the *stepwise* formation of the complex ions, e.g.,

$$Cu(NH_3)^{++} + NH_3 \rightleftarrows Cu(NH_3)_2^{++}$$

and stepwise formation constants.

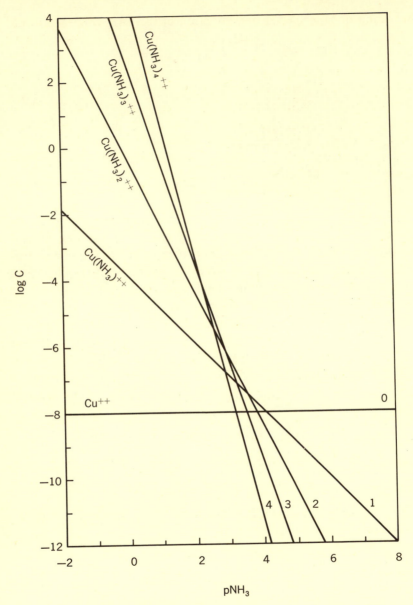

FIGURE 8.9. Master variable diagram for the copper (II)-ammonia complex ion system.

which species predominate at each $[NH_3]$. The 4-coordinated species, $Cu(NH_3)_4^{++}$, is dominant at pNH_3's below 2, or at ammonia concentrations greater than 0.01 M, although there are appreciable amounts of the 3-coordinated species present up to ammonia concentrations of 1 M. At ammonia concentrations less than 10^{-4} M, the uncomplexed species predominates.

The reader might now question these results, since they are based on the arbitrary assumption that the $[Cu^{++}]$ was 10^{-8} M. What if it was actually 10^{-12} M? Since changes in the $[Cu^{++}]$ (i.e., the 0 line) will shift all of the lines up or down by the same amount (as if lines 0, 1, 2, 3, and 4 are rigidly attached together and can slide vertically up and down the coordinate axes), the relative concentrations of the different species will be unchanged by changes in the $[Cu^{++}]$. Therefore one need only choose a convenient concentration for these calculations.

DISTRIBUTION DIAGRAMS

Using the master variable diagram, we can plot a distribution diagram, showing the fraction of the total copper, C_{Total}, present as each species, at a given $[NH_3]$. For example, the fraction of total copper present as $Cu(NH_3)_4^{++}$, β_4, is given by

$$\beta_4 = \frac{[Cu(NH_3)_4^{++}]}{C_{Total}} \tag{8.43}$$

Referring to Figure 8.9, we see that at a $pNH_3 = 0$,

$$[Cu(NH_3)_4^{++}] = 3.9 \times 10^4 \ M$$

$$[Cu(NH_3)_3^{++}] = 3.0 \times 10^2 \ M$$

$$[Cu(NH_3)_2^{++}] = 0.4 \ M$$

$$[Cu(NH_3)^{++}] = 1.4 \times 10^{-4} \ M$$

$$[Cu^{++}] = 10^{-8} \ M$$

$$C_{Total} = [Cu(NH_3)_4^{++}] + [Cu(NH_3)_3^{++}] + \cdots [Cu^{++}]$$

$$C_{Total} = 3.93 \times 10^4 \ M$$

Therefore, at a $pNH_3 = 0$,

$$\beta_4 = \frac{3.9 \times 10^4}{C_{Total}} = 0.992$$

$$\beta_3 = \frac{3.0 \times 10^2}{C_{Total}} = 0.008$$

down to

$$\beta_0 = \frac{10^{-8}}{C_{Total}} = 2.5 \times 10^{-13}$$

Calculations such as these at different pNH_3's allow construction of the distribution diagram shown in Figure 8.10.

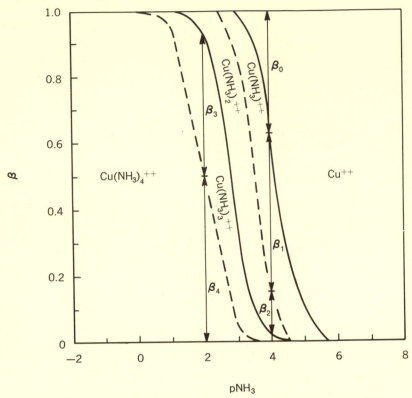

FIGURE 8.10. Distribution diagram for the copper(II)-ammonia complex ion system. The fraction of the total copper(II) present as the noncomplexed, and the 1, 2, 3, or 4,-coordinated species, β_0, β_1, β_2, β_3, and β_4, respectively, are given by the distances between the boundary lines.

$\bar{n}$ DIAGRAM

One other type of diagram which is used frequently by those who work in the field of complex ion chemistry is the $\bar{n}$ diagram. $\bar{n}$ is called the *average ligand number* and is defined as the average number of ligand groups attached to the central metal ion at a given concentration of ligand.* For example, referring to Figures 8.9 or 8.10, we see that at a pNH_3 of 7 the average ligand number is about 0, while at a pNH_3

---

* $\bar{n}$ is sometimes called *Bjerrum's formation function* after one of the early workers in the field of complex ion chemistry who appreciated the importance of stepwise formation of complex ions.

of 1, $\bar{n}$ is close to 4. At other pNH$_3$ values we can calculate $\bar{n}$ from the distribution diagram. We can calculate $\bar{n}$ from the equation

$$\bar{n} = (0)\beta_0 + (1)\beta_1 + (2)\beta_2 + (3)\beta_3 + (4)\beta_4 \qquad (8.44)$$

We look up the requisite β's on the distribution diagram and then calculate $\bar{n}$. For example at a pNH$_3$ of 3, we find that $\beta_4 = 0.05$, $\beta_3 = 0.27$, $\beta_2 = 0.53$, $\beta_1 = 0.14$, and $\beta_0 = 0.01$. Therefore

$$\bar{n} = (0)(0.01) + (1)(0.14) + 2(0.53) + 3(0.27) + 4(0.05)$$

$$\bar{n} = 2.11$$

Using calculations such as these, at different pNH$_3$'s, Figure 8.11 was constructed.

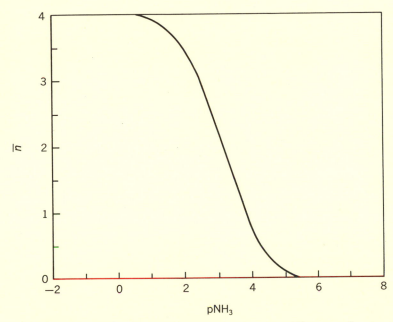

FIGURE 8.11. $\bar{n}$ diagram for the copper(II)-ammonia complex ion system.

$\bar{n}$ diagrams are useful in experimentally determining the formation constants of complex ions. The average ligand number is just the total concentration of ligand coordinated with the metal divided by the total amount of metal ion added to the solution. For example, if we add C_L moles per liter of NH$_3$ to a solution, and add C_M moles per liter of copper, and we find that the free (uncoordinated) ammonia concentration in the resulting solution is [L], then the total number of moles of ammonia complexed is $C_L-[L]$, and the average ligand number is

$$\bar{n} = \frac{C_L-[L]}{C_M} \qquad (8.45)$$

Experimentally, we know how many moles of metal and ligand we add to the solution and the problem of finding $\bar{n}$ is resolved into finding the concentration of uncomplexed ligand, $[L]$, which can be done in a variety of methods. An $\bar{n}$ diagram can then be constructed experimentally. The problem of the investigator is to calculate the stability constants for the various complex ion species from this $\bar{n}$ diagram.

8–6. CONCLUSIONS

Graphical methods are useful for solving equilibrium problems, for enabling the visualization of conditions in a solution, and for representing a system over a range of conditions. Another type of diagram that is particularly useful for representing oxidation-reduction systems, is the *predominance area* diagram. In an oxidation-reduction system, different species can be present, depending upon the potential and the pH of the solution. For example, in the lead system possible species include Pb, Pb^{++}, $Pb(OH)_2$, PbO_2, $HPbO_2$, etc. A *potential-pH diagram* can aid in visualizing this kind of system under a range of conditions. The interested reader is referred to several discussions of this kind of diagram in the literature.

PROBLEMS

8.1. Construct a log C vs. pH diagram for the following acid-base systems.

(a) $0.10\ M\ HNO_2$ (b) $0.10\ M\ H_2CO_3$
(c) $0.10\ M\ H_3PO_4$ (d) $0.20\ M\ NH_4Cl$

8.2. Using the diagrams constructed in Problem 8.1, calculate the pH and the concentration of all species, in the following solutions.

(a) $0.10\ M\ KNO_2$ (e) $0.10\ M\ (NH_4)_2CO_3$
(b) $0.05\ M\ KNO_2$ and   (f) $0.10\ M\ Na_2CO_3$
    $0.05\ M\ HNO_2$ (g) $0.10\ M\ NaH_2PO_4$
(c) $0.10\ M\ NaHCO_3$ (h) $0.10\ M\ Na_2HPO_4$
(d) $0.10\ M\ NaHCO_3$ and   (i) $0.10\ M\ Na_3PO_4$
    $0.20\ M\ NH_4Cl$ (j) $0.10\ M\ (NH_4)_2PO_4$

(k) 0.10 M Na_3PO_4 and
      0.20 M NH_4Cl

(l) 0.10 M NaH_2PO_4 and
      0.20 M NH_4Cl

8.3. Draw distribution diagrams for the systems in Problem 8.1.

8.4. Using the K_{sp}'s of the following, Bi_2S_3, FeS, NiS, CdS, PbS, ZnS, CoS, MnS; K_1 and K_2 for H_2S; assuming the solution is saturated in H_2S ($[H_2S] = 0.10$ M)

(a) Plot a log C vs. pH diagram for pH's of -2 to 9 for the above metals in a solution saturated in H_2S.

(b) Use this diagram to find:
  (1) the pH range for the separation of Fe^{++} from Mn^{++} if both are present at a 0.01 M concentration.
  (2) the pH range for the separation of Fe^{++} and Mn^{++} if the $[Fe^{++}] = 0.01$ M and the $[Mn^{++}] = 0.1$ M.
  (3) which of the following separations can be made quantitatively: (a) Pb^{++} from Cd^{++} (0.1 M solutions); (b) Co^{++} from Zn^{++} (0.1 M solutions); (c) Zn^{++} from Fe^{++} (0.01 M solutions)

(c) Separation of Group II metals from Group III metals in the usual qualitative analysis scheme is carried out in 0.3 M HCl. If all the metal ions are assumed present at a 0.01 M concentration, which metals will remain unprecipitated when the solution is made 0.3 M in HCl and saturated with H_2S?

8.5. When a Zn^{++} solution is treated with NH_3, a series of Zn^{++}—NH_3 complexes are formed:

$$Zn^{++} + NH_3 \rightleftarrows Zn(NH_3)^{++} \qquad \log K_1 = 2.27$$
$$Zn^{++} + 2NH_3 \rightleftarrows Zn(NH_3)_2^{++} \qquad \log K_2 = 4.61$$
$$Zn^{++} + 3NH_3 \rightleftarrows Zn(NH_3)_3^{++} \qquad \log K_3 = 7.01$$
$$Zn^{++} + 4NH_3 \rightleftarrows Zn(NH_3)_4^{++} \qquad \log K_4 = 9.06$$

(a) Plot a log C vs. pNH_3 diagram for this system. (Take $\log [Zn^{++}]$ $= -8.00$, plot log C from 1 to -12 and pNH_3 from -2 to 10).
(b) Plot a distribution diagram for this system.
(c) Plot an $\bar{n}$ diagram for this system.
(d) Plot β_0 vs. pNH_3.

8.6. Draw a log C vs. $pCrO_4^{--}$ diagram for Ag_2CrO_4.
(a) From the diagram, calculate the solubility of Ag_2CrO_4 in water (neglect the basicity of chromate ion).

(b) Add lines to the diagram for $HCrO_4^-$ at pH's of 1.0, 2.0, 4.0, and 7.0.

(c) Calculate the concentrations of the various species and the solubility of Ag_2CrO_4 in solutions in equilibrium with solid Ag_2CrO_4 at pH's of 1.0, 2.0, 4.0, and 7.0.

8.7. Draw a log C vs. pS^{--} diagram for CuS.

(a) From the diagram calculate the solubility of CuS in water (neglecting the basicity of sulfide ion).

(b) Add lines for H_2S and HS^- at pH's of 0, 1.0, 2.0, and 4.0.

(c) Calculate the solubility of CuS in solutions of pH 0, 1.0, 2.0, and 4.0.

SUPPLEMENTARY READING

Boyd, R. N., Procedure for Solving Equilibrium Problems, *J. Chem. Educ.*, **29**, 198 (1952). Solving problems by the "lumped equilibrium constant" method.

Davidson, D., and K. Geller, Algebra of Simultaneous Equilibria, *J. Chem. Educ.*, **30**, 238 (1953). "Lumped equilibrium constant" methods. See also letters, pp. 532–534.

Delahay, P., Pourbaix, M., and P. Van Rysselberghe, Potential-pH Diagrams, *J. Chem. Educ.*, **27**, 683 (1950).

Freiser, H., and Q. Fernando, Teaching Ionic Equilibrium. Use of Log-Chart Transparencies, *J. Chem. Educ.*, **42**, 35 (1965). Discusses master-variable graphical techniques in acid-base calculations and the use of commercially available transparencies for construction of these graphs.

Fritz, J. J., Systematic Calculation of Ionic Equilibria, *J. Chem. Educ.*, **30**, 442 (1953). Solving problems by the "principle reaction" method.

Guerra, J., Application of Logarithmic Triangular Charts to Ionization Equilibria of Weak Acids, *J. Chem. Educ.*, **34**, 341 (1957). An alternate method of treating acid-base equilibria graphically.

Sillen, L. G., Oxidation-Reduction Diagrams, *J. Chem. Educ.*, **29**, 600 (1952). Master variable diagrams of redox systems.

CHAPTER 9 SEPARATIONS AND OTHER EQUILIBRIA

9–1. *INTRODUCTION*

The application of equilibrium theory to various classes of chemical reactions has been discussed. In this chapter we consider some different types of problems, including those involving separations, which are of general interest to chemists. Separations are used in all branches of chemistry. The analytical chemist often faces the problem of separating the different components of a sample before he can determine how much of each is present. The chemist investigating the structure of natural products must first isolate the desired compound from the host of other substances that are found with it in nature. A chemist synthesizing a compound must remove the desired product from a reaction mixture containing other substances. Finally, most industrial processes involve separations: obtaining the desired constituent from raw materials or separating reaction mixtures into their various components.

Usually a separation method depends upon the formation of two distinct physical phases, such as a solid and a liquid, a gas and a liquid, or two immiscible liquids, where the desired constituent is in one phase and the undesired constituent(s) in the other. Finally the two phases are physically separated. One method of separation, by precipitation, has already been discussed. In this separation the desired constituent is made to form an insoluble substance (precipitate), while the other constituents remain in solution. The physical method of separation is accomplished by filtration; the precipitate remains behind on the filter paper with the desired constituent, and the solution containing the other material passes through. Separations can also be accomplished by extraction, ion exchange, electrolysis, and a number of other techniques.

9–2. LIQUID-LIQUID EXTRACTION

THE DISTRIBUTION LAW

Extraction methods involve the distribution of a substance between two immiscible liquids. Immiscible liquids are those which do not dissolve in one another; when they are poured together they do not form a single solution, as alcohol and water do, but form two distinct layers. Examples of immiscible liquids are water and carbon tetrachloride, water and oil, and water and mercury. If a substance is equilibrated with two immiscible liquids, it will distribute itself between the two according to its relative solubility in each. If it is much more soluble in one, most of it will go into that layer, and can be removed simply by separating that layer which contains it.

Consider shaking up a mixture of carbon disulfide and water, which are immiscible, and some iodine. Iodine is about 400 times more soluble in carbon disulfide than in water, and so the I_2 will mostly go into the carbon disulfide layer. The extent to which a substance is extracted can be found by using equilibrium theory. The distribution of a substance X between two phases, o and w can be represented by the equilibrium

$$X(w) \rightleftarrows X(o) \tag{9.1}$$

so that the equilibrium constant expression for this distribution (sometimes called the *Nernst Distribution Law*) is:

$$\frac{[X]_o}{[X]_w} = K_D \tag{9.2}$$

where $[X]_o$ is the concentration* of X in liquid o and $[X]_w$ is the con-

---

* As in most other equilibrium constant expressions, activites rather than concentrations should be used for nonideal solutions.

centration of X in liquid *w*. K_D is called the *distribution coefficient* (or sometimes the *partition coefficient*). The larger the distribution coefficient is, the more of the substance which will go into phase *o* and the more complete the separation will be. The principles of extraction separations can be illustrated by some sample problems.

Example 9.1. For the distribution of I_2 between carbon disulfide and water, $K_D = 410$, with I_2 more soluble in carbon disulfide (CS_2). 100 ml of a 0.010 *M* aqueous iodine solution is shaken with 10 ml of CS_2. What is the concentration of iodine in the CS_2 layer at equilibrium? What percentage of the total iodine is extracted into the CS_2?

Let x = number of moles of I_2 left in the water layer after extraction with CS_2. Then (0.1 liter (0.01 mole/liter) − x) mole is extracted into CS_2.

$$\frac{[I_2]_{CS_2}}{[I_2]_w} = 410 = \frac{0.001 - x \text{ mole}/0.01 \text{ liter}}{x \text{ mole}/0.1 \text{ liter}}$$

$$\frac{(0.001 - x)}{(x)} = 41$$

$$x = 2.4 \times 10^{-5} \text{ mole}$$

Concentration in CS_2 layer is

$$\frac{(0.001 - 2.4 \times 10^{-5}) \text{ mole}}{0.010 \text{ liter}} = 0.098 \ M \quad Answer$$

$$\% \text{ extracted} = \frac{0.98 \times 10^{-3} \text{ mole}}{0.0010 \text{ mole}} \times 100 = 98\% \quad Answer$$

For this system, one extraction with carbon disulfide removes most of the iodine from the water layer. The net amount of material extracted depends upon the amount of extracting phase used.

Example 9.2. 100 ml of a 0.010 *M* iodine solution are shaken with 5 ml of CS_2. What is the concentration of I_2 in the CS_2 layer and what percentage of the iodine is extracted from the water?

The same expressions as in Example 9.1 apply, except the concentration in the CS_2 layer is

$$[I_2]_{CS_2} = \frac{(0.001 - x) \text{ mole}}{0.005 \text{ liter}}$$

$$\frac{(0.001 - x) \text{ mole}/0.005 \text{ liter}}{x \text{ mole}/0.1 \text{ liter}} = 410$$

$$\frac{0.001 - x}{x} = 20.5$$

$$x = 4.6 \times 10^{-5} \text{ mole} \quad \text{(left in water)}$$

$$[I_2]_{CS_2} = \frac{0.954 \times 10^{-3} \text{ mole}}{0.005 \text{ liter}} = 0.19 \text{ } M \quad \text{Answer}$$

$$\% \text{ extracted} = \frac{0.95 \times 10^{-3} \text{ mole}}{0.0010 \text{ mole}} \times 100 = 95\% \quad \text{Answer}$$

Less iodine is extracted when the volume of the extracting phase is decreased.

Example 9.3. The water phase from the above is extracted with an additional 5 ml of CS_2. How many moles of I_2 are extracted? What total percentage of iodine is extracted for the two 5 ml portions?

If $y =$ amount of I_2 left in water, the concentration in the CS_2 phase is $\dfrac{(4.6 \times 10^{-5} - y) \text{ mole}}{0.005 \text{ liter}}$ and

$$\frac{(4.6 \times 10^{-5} - y)/0.005}{y/0.100} = 410$$

$$y = 2.1 \times 10^{-6} \text{ mole} \quad \text{(left in water)}$$

Then the moles extracted into the second 5 ml volume of CS_2 is 4.4×10^{-5} mole. The total moles of iodine extracted in the two extractions is 0.999×10^{-3} mole, or the total extracted is 99.9 percent. Notice that more is extracted with two 5 ml portions than with one 10 ml portion.

The principle of repeated extractions is even more important when the distribution coefficient is smaller.

COMPLEX ION EQUILIBRIA

Extraction methods can also be applied to the separation of metal ions. The species containing the metal ion which is extracted into an organic phase from the aqueous phase must be in the form of a neutral species, since ions do not exist in most of the organic liquids used as extractants, such as chloroform, carbon tetrachloride, or benzene. The species may sometimes be a simple complex ion. A frequent method of separating iron(III) from many other metal ions involves extracting a 6 M HCl solution of the ions with diethyl ether. The iron in this solution is in the form of the neutral complex $HFeCl_4$, which is quite soluble in the ether layer. Many other metal ions are in the form of charged species in this solution and do not extract into the ether layer.

A useful method for separating metal ions involves the extraction of neutral chelate complexes (see Chapter 5) of metal ions into organic liquids. In this technique an aqueous solution of the metal ion is shaken with an organic liquid containing the chelating agent. The chelating agent, HX, is distributed between the water and the organic phase, and causes formation of the metal chelate, MX_n, which extracts to some extent into the organic phase. For example 8-hydroxyquinoline (sometimes called oxine) is a bidentate chelating agent with the structure

The singly ionized form of 8-hydroxyquinoline will react with many metal ions to form an uncharged species which can be extracted into chloroform. For example, with Mg^{++} the species formed is

Equilibrium calculations of these extractions can become quite complicated because a number of different equilibria are involved. For example, the distribution of both the chelating agent, HX, and the metal chelate, MX_n, between water and the organic phase must be considered:

$$(HX)_w \rightleftarrows (HX)_o \qquad [HX]_o/[HX]_w = K_{D,HX} \qquad (9.3)$$

$$(MX_n)_w \rightleftarrows (MX_n)_o \qquad [MX_n]_o/[MX_n]_w = K_{D,MXn} \qquad (9.4)$$

Moreover the stability constant of the metal chelate must be considered

$$M^{+n} + nX^- \rightleftarrows MX_n \qquad [MX_n]_w/[M^{+n}]_w[X^-]_w^n = K_{stab} \qquad (9.5)$$

as well as the acidity constant of the chelating agent

$$HX \rightleftarrows H^+ + X^- \qquad [H^+]_w[X^-]_w/[HX]_w = K_a \qquad (9.6)$$

The dissociations of HX and MX_n in the organic phase are usually not considered, since ionization does not occur to any great extent in the organic liquids usually used. Qualitatively, the extent of extraction of a given metal species will depend upon the concentration of the chelating agent, the pH, and the magnitude of the various equilibrium constants governing the extraction. When considering the separation of two different metals, both from extractable chelates, it is often possible to

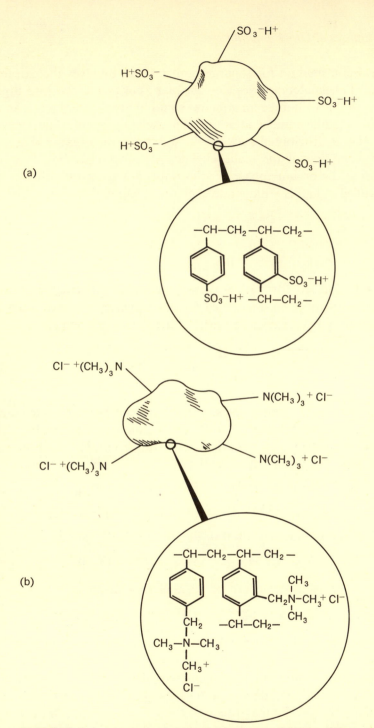

FIGURE 9.1. Representation of (a) cation exchange resin and (b) anion exchange resin.

calculate conditions in which one metal ion will extract while another will not. Although further calculations of this type are beyond the scope of our treatment, the interested reader can find several references that contain detailed calculations of this type and give numerous examples of practical separation schemes (see also Problem 9.4).

9–3. ION EXCHANGE

ION EXCHANGE EQUILIBRIA

Ion exchange equilibria are examples of solid-solution reactions. An *ion exchange resin* is an organic network containing groups that can attract and hold cations or anions to them. For example, a cation exchange resin can be represented by $Res—SO_3^- H^+$, where Res is the organic structure of the resin, which can be thought of as providing a support for the active $—SO_3^-$ sites. The $—SO_3^-$ sites can attract cations and hold them on the surface of the solid resin. Figure 9.1(a) represents a particle of cation exchange resin containing H^+. Suppose cation exchange resin in its acidic form (that is, containing H^+ on the $—SO_3^-$ sites) is mixed with a solution containing Na^+ ions. The Na^+ ions will exchange to some extent with H^+ and an ion exchange equilibrium will be established.

$$Res—SO_3^- H^+ + Na^+ \rightleftarrows Res—SO_3^- Na^+ + H^+ \qquad (9.7)$$

In general the exchange of ions A^+ and B^+, with the cation exchange resin represented by R^-, is given by

$$R^-A^+ + B^+ \rightleftarrows R^-B^+ + A^+ \qquad (9.8)$$

and yields the following equilibrium constant expression

$$\frac{a_{A^+} a_{R-B^+}}{a_{B^+} a_{R-A^+}} = K_A^{\ B} \qquad (9.9)$$

where the a's represent activities of the species in the solution and the resin. If we replace activities by concentrations in the solution phase (Chapter 7) and by mole fractions in the resin phase, (9.9) becomes

$$\frac{[A^+] X_{B^+ (resin)}}{[B^+] X_{A^+ (resin)}} = K'_A{}^B \qquad (9.10)$$

where the activity coefficients have been lumped into $K_A^{\ B}$ to yield $K'_A{}^B$. The mole fraction X_{A^+} represents the number of moles (or millimoles) of A^+ in the resin, divided by the total number of moles (or millimoles) of exchangeable cation in the resin. If

$$n_{A^+} = \text{millimoles of } A^+ \text{ in resin}$$

$$n_{B^+} = \text{millimoles of } B^+ \text{ in resin}$$

$$X_{A^+} = n_{A^+}/(n_{A^+} + n_{B^+}) \qquad X_{B^+} = n_{B^+}/(n_{A^+} + n_{B^+})$$

(9.10) can be written

$$\frac{[A^+]n_{B^+}}{[B^+]n_{A^+}} = K'^{B}_{A} \qquad (9.11)$$

The resin phase is in many ways like a concentrated electrolyte solution, so that K'^{B}_{A} may be very different for different X_{A^+}/X_{B^+} ratios, and can only be considered constant under conditions where only small changes in the composition of the resins occur. For the exchange of unequally charged species, such as

$$R^- Na^+ + \tfrac{1}{2}Ca^{++} \rightleftarrows \tfrac{1}{2}R_2^{--} \cdot Ca^{++} + Na^+ \qquad (9.12)$$

The following equilibrium constant expression results:

$$\frac{[Na^+]X^{\frac{1}{2}}_{Ca^{++}(resin)}}{[Ca^{++}]^{\frac{1}{2}}X_{Na^+(resin)}} = K'^{Ca^{++}}_{Na^+} \qquad (9.13)$$

An anion exchange resin contains fixed active cation groups, such as $-N(CH_3)_3^+$, which holds anions to them (Fig. 9.1(b)). The following equation can be written for the exchange of anions X^- and Y^- (representing the resin by R^+)

$$R^+X^- + Y^- \rightleftarrows R^+Y^- + X^- \qquad (9.14)$$

Expressions similar to (9.9), (9.10), and (9.11) can be written for this reaction. Typical values of K'^{B}_{A} for both cation and anion exchange resins are given in Table 9.1.

Exchange resins have a certain *capacity*, represented by the number of replaceable millimoles of univalent ion (milliequivalents) per gram of resin. Typically the capacity of a resin would be 4 to 5 milliequivalents (meq) per gram.

Example 9.4. 100 ml of a solution containing 10^{-3} moles/liter of $AgNO_3$ is treated with 1.0 gram of an 8 percent crosslinked cation exchange resin in the H^+-form with a capacity of 5.0 meq/g. Calculate the concentration of Ag^+ remaining in the solution and the amount of Ag^+ on the resin.

TABLE 9.1   Table of $K_A{}^B$ for Ion Exchange Resins

Replaceable Ion	$K_A{}^B$ for crosslinking of		
	4%	8%	16%
Cations			
(Univalent)			
Li	1.00	1.0	1.0
H	1.32	1.27	1.47
Na	1.58	1.98	2.37
K	2.27	2.90	4.50
Rb	2.46	3.16	4.62
Cs	2.67	3.25	4.66
Ag	4.73	8.51	22.9
Tl	6.71	12.4	28.5
(Divalent)			
Mg	2.95	3.29	3.51
Ca	4.15	5.16	7.27
Sr	4.70	6.51	10.1
Ba	7.47	11.5	20.8
Pb	6.56	9.91	18.0
	2%	8%	
Anions			
OH	0.80	0.50	
F		0.08	
Cl	1.0	1.0	
Br	2.7	3.5	
I	9.0	18.0	
NO_3		3.0	
SCN	6.0	4.3	
ClO_4	9.0	10.0	

NOTE: Cation exchange resin constants are referred to Li^+; anion exchange constants to Cl^-. To calculate $K_A{}^B$ from $K_{Li}{}^A$ and $K_{Li}{}^B$, use the formula, $K_A{}^B = K_{Li}{}^B / K_{Li}{}^A$. Units for univalent-divalent exchanges are mole fractions of cations in resins, normality (meq/ml) in solution, with K expressed with divalent ion concentration to the one-half power (see (9.13)). Data are for polystyrene base resins, sulfonic acid cation exchanger, and type 2 quaternary base anion exchanger.

From *Handbook of Analytical Chemistry*, by L. Meites. Copyright © 1963 by McGraw-Hill Book Company. Used by permission of McGraw-Hill Book Company.

The exchange reaction is

$$R^-H^+ + Ag^+ \rightleftarrows R^-Ag^+ + H^+ \tag{9.15}$$

$$\frac{[H^+]n_{Ag^+}}{[Ag^+]n_{H^+}} = K_{H^+}{}^{Ag^+} \tag{9.16}$$

From Table 9.1

$$K_{H^+}^{Ag^+} = \frac{K_{Li^+}^{Ag^+}}{K_{Li^+}^{H^+}} = \frac{8.51}{1.27} = 6.70$$

Since the capacity of the resin is 5.0 meq/g, the total millimoles of univalent cation on the resin is given by

$$n_{Ag^+} + n_{H^+} = 5.0 \text{ mmol}$$

In solution, by electroneutrality,

$$[H^+] + [Ag^+] = [NO_3^-] = 10^{-3} \tag{9.17}$$

Let $x = [Ag^+]$; then

$$n_{Ag^+} = 100(10^{-3} - x) \text{ mmol}$$
$$n_{H^+} = 5.0 - 100(10^{-3} - x) \text{ mmol}$$
$$[H^+] = 10^{-3} - x \ M$$

A direct solution of the problem involves introducing the above 4 values into (9.16) and solving for x. If we assume $x \ll 10^{-3}$ then

$$n_{Ag^+} = 0.1 \text{ mmol}$$
$$n_{H^+} = 4.9 \text{ mmol}$$
$$[H^+] = 10^{-3} \ M$$

and

$$[Ag^+] = x = \frac{(10^{-3})(0.1)}{(4.9)(6.7)} = 3.0 \times 10^{-6} \ M \quad \textit{Answer}$$

millimoles Ag^+ on resin $= 100 \, (10^{-3} - 3 \times 10^{-6}) \approx 0.10$ millimole.
Example 9.5. Consider the same problem as above for a solution containing 0.1 M HNO$_3$. The equations are the same, except

$$[H^+] = 0.1 + 10^{-3} - x$$

Since the $[Ag^+]$ is so much smaller than the $[H^+]$, the $[H^+]$ is essentially constant and equal to 0.1 M. Similarly, since the maximum value of n_{Ag^+} is 0.1, n_{H^+} is practically constant and equal to 5.0 (actually 4.9 < n_{H^+} < 5.0). Therefore

$$\frac{n_{Ag^+}}{[Ag^+]} \approx \frac{6.7 \, (5.0)}{0.1} \approx 3.3 \times 10^2$$

or, letting $x = [Ag^+]$

$$100(10^{-3} - x) = 330x$$

$$[Ag^+] = x = 10^{-1}/430 = 2.3 \times 10^{-4} \ M$$

DISTRIBUTION COEFFICIENTS

The conditions in Example 9.5 are typical of many ion exchange experiments; i.e., the total amount of the ion of interest is small compared to the capacity of the resin, and the concentration of the exchangeable ion, e.g., H^+, is high or fixed in the solution. For example, considering the process given in (9.9), when the $[A^+] \gg [B^+]$, and $n_{A^+} \gg n_{B^+}$, both $[A^+]$ and n_{A^+} can be taken as constants, and (9.11) can be rearranged to

$$\frac{n_{B^+}}{[B^+]} = D \qquad (9.18)$$

where D, the *distribution coefficient*, is a function of n_{A^+}, and $[A^+]$. Frequently D values are tabulated as

$$D = \frac{\text{amount of B/gram resin}}{\text{amount of B/ml solution}} \qquad (9.19)$$

Note that (9.18) and the equation governing extraction equilibria, (9.2), are very similar, so that under the conditions at which (9.18) holds, ion exchange can be considered the equilibrium distribution of B between an aqueous phase and the resin phase

$$B^+_{(w)} \rightleftarrows B^+_{(resin)} \qquad (9.20)$$

EFFECT OF COMPLEXATION

If B^+ is a metal ion in the presence of an excess of complexing agent, (X^-), it will also be involved in an equilibrium involving complexation of B^+ in the solution

$$B^+ + pX^- \rightleftarrows BX_p^{-p+1} \qquad (9.21)$$

An equation like (9.19) can still be written for the exchange of the metal species on a resin, but B^+ will be in solution as BX_p^{-p+1} and D will contain the stability constant of the complex ion and be a function of the concentration of the complexing agent.

A method of separating metal ions is based on their behavior on an anion exchange resin in HCl solutions, where chloro-complexes of

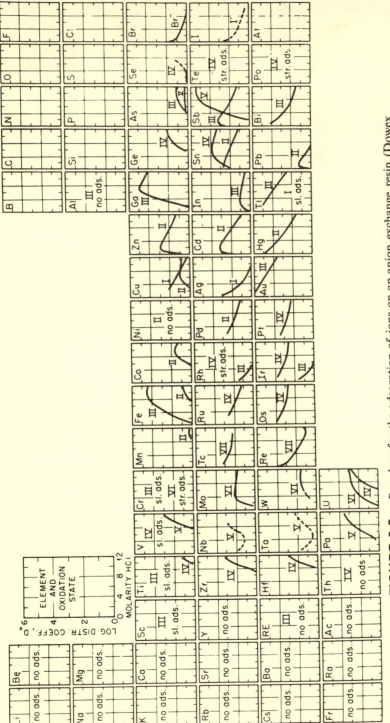

FIGURE 9.2. D_v-values for the adsorption of ions on an anion exchange resin (Dowex 1-X10) from hydrochloric acid: no ads.—no adsorption, $0.1 M < M$HC1 $< 12 M$; sl. ads.— slight adsorption in $12 M$ HCl ($0.3 < D_v < 1$); str. ads.—strong adsorption ($D_v \gg 1$). D_v is amount per ml resin/amt per ml solution, or since the density of the resin is about 0.5 g/cm³, D (amt per g resin/amt per ml soln) = 0.5 D_v. From I. M. Kolthoff and P. J. Elving (eds.), *Treatise on Analytical Chemistry*, Part 1, Vol. 3, New York: Interscience, 1959. By permission of John Wiley & Sons, Inc.

different stabilities are formed. The differences in stability of the complexes and the different concentrations of H^+ are reflected in a D for each metal that varies with the HCl concentration (Fig. 9.2).

Example 9.6. When 10 ml of 10^{-2} M Fe^{+++} solution in 6 M HCl is treated with 1.0 g of an anion exchange resin, Dowex 1-X10, what percentage of the Fe^{+++} is left in solution?

From Figure 9.2, $D = 10^3$, so that, if $x = $ final concentration of iron(III)

$$\frac{\text{mmol iron(III)/g resin}}{\text{mmol iron(III)/ml soln}} = 10^3 = \frac{10(10^{-2} - x)}{x}$$

$$x = 10^{-4} M$$

$$\% \text{ iron(III) left in soln} = \frac{10^{-4}}{10^{-2}} \times 100 = 1\% \quad \textit{Answer}$$

Example 9.7. If the solution resulting from the above treatment is treated with an additional 1.0 gram of resin, what percentage of the original Fe^{+++} will remain?

If $x' = $ final concentration of iron after 2nd addition,

$$\frac{10(10^{-4} - x')}{x'} = 10^3$$

$$x' = 10^{-6}$$

$$\% \text{ iron(III) left in solution} = \frac{10^{-6}}{10^{-2}} \times 100 = 0.01\% \quad \textit{Answer}$$

Note that two treatments with resin are needed in this case to reduce the concentration to below 0.1 percent. For lower D values, many repeated additions of resin may be necessary, and for this reason ion exchange is often carried out using a column technique. The resin (10 to 15 grams) is packed in a column and the solution is poured through. As the solution passes down the column it undergoes repeated ion exchange steps, so that passage through a column is equivalent to a number of individual (batch) experiments. Each effective equilibration step on the column is called a *theoretical plate*. A column which carries out a separation that would require ten repeated batch steps is said to have ten theoretical plates, under the given flow and packing conditions.

9–4. ELECTRODEPOSITION

A useful method of separating metal ions from a solution is through their electrolytic deposition on a solid (e.g., platinum) or mercury electrode. The apparatus for carrying out an electrodeposition is shown in Figure 9.3 and usually consists of a dc power supply providing the electrolysis voltage and a means of measuring the potential

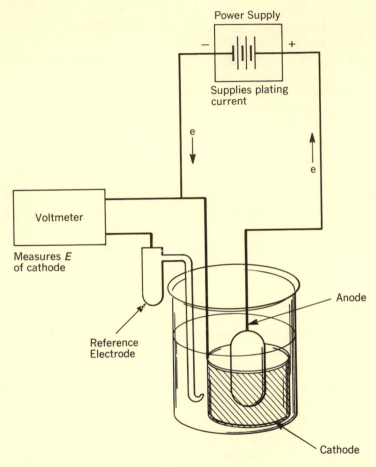

FIGURE 9.3. Apparatus for controlled potential electrodeposition.

of the electrode, E, with respect to some stable reference electrode. The electrodeposition equilibrium is just the half-reaction of the redox couple of interest, such as

$$Cu^{++} + 2e \rightleftarrows Cu \qquad (9.22)$$

or in general

$$O + ne \rightleftarrows R \qquad (9.23)$$

The equilibrium constant expression which governs the equilibrium is

$$\frac{[R]}{[O]} = K_E = 10^{-\frac{n(E-E^0)}{0.059}} \tag{9.24}$$

or in logarithmic form (called the *Nernst equation*)

$$E = E^0 - \frac{0.059}{n} \log \frac{[R]}{[O]} \tag{9.25}$$

where E is the electrode potential of the electrode where electrodeposition is occurring with respect to the reference electrode in volts, and E^0 is the standard potential of the half-reaction with respect to the same reference electrode (Appendix D). Note that (9.24) is the half-reaction equivalent of (6.7) and is then analogous to the other equations governing distribution between phases, such as (9.2) and (9.18), except that the distribution constant, K_E, is a function of electrode potential. This dependence on potential arises from the presence of electrons as reactants. When a metal is plated on a solid electrode, R is a solid at unit activity.* When a mercury electrode is used and R is soluble in the mercury, forming an amalgam, $[R]$ is the concentration in the mercury electrode, i.e., mmol of R per ml of mercury.

Example 9.8. 100 ml of a 10^{-2} M Pb^{++} solution is electrolyzed with a platinum cathode maintained at a potential of -0.244 V.

(a) Calculate the concentration of Pb^{++} remaining in solution after completion of electrolysis.

$$Pb^{++} + 2e \rightleftarrows Pb \quad E^0 = -0.126 \ V$$

$$\frac{1}{[Pb^{++}]} = K_E = 10^{\frac{-2[-0.244-(-0.126)]}{0.059}}$$

Since Pb is a solid, its activity is taken as one. Then

$$[Pb^{++}] = 10^{-4} \ M$$

(b) Calculate the number of moles of Pb deposited.

Amount originally in solution $= 100 \ ml \times 10^{-2} \ mmol/ml$

$$= 1.00 \ mmol$$

Amount left after deposition $= 100 \ ml \times 10^{-4} \ mmol/ml$

$$= 0.01 \ mmol$$

Amount deposited $= 1.00 - 0.01 = 0.99 \ mmol$

* If less than a monolayer of metal is plated on an electrode, the activity of R may be less than unity. The assumption of $[R] = 1$ cannot be used when the plating of small amounts of material from very dilute solutions is considered.

Example 9.9. Repeat Example 9.8 for deposition on a mercury cathode with a volume of 10.0 ml.

When Pb is plated on mercury, it dissolves to form lead amalgam

$$Pb^{++} + Hg + 2e \rightleftarrows Pb(Hg)$$

Assuming the potential for this reaction is the same as above*

$$\frac{[Pb]_{Hg}}{[Pb^{++}]} = 10^4$$

Let $x = $ mmol of Pb^{++} left in solution after plating. Then $1.00 - x = $ mmol of Pb in mercury

$$\frac{\dfrac{1.00 - x}{10}}{\dfrac{x}{100}} = 10^4$$

$$x = 1.00 \times 10^{-3} \text{ mmol}$$

$$[Pb^{++}] = \frac{x}{100} = 1.00 \times 10^{-5} \ M$$

Amount deposited $= 1.00 - 10^{-3} \approx 1.00$ mmol

A more complete deposition results from making the potential more negative. Selectivity for different metals is accomplished by adjusting the potential to a value at which one metal will deposit and another will not.

Example 9.10. A solution contains $1.00 \times 10^{-4} \ M$ Ag^+ and $1.00 \times 10^{-2} \ M$ Cu^{++} in 1 M HNO_3. Over what range of potentials can a quantitative separation of Ag and Cu be accomplished by plating silver, leaving Cu^{++} unplated in solution?

$$Ag^+ + e \rightleftarrows Ag \quad E^0 = +0.80 \ V$$

$$Cu^{++} + 2e \rightleftarrows Cu \quad E^0 = +0.34 \ V$$

For quantitative plating of Ag, the $[Ag^+]$ must be reduced by a factor of 10^3, so that, after electrolysis

$$[Ag^+] = 10^{-3} \times 10^{-4} = 10^{-7} \ M$$

---

* Actually the potential for plating a metal into mercury differs somewhat from the standard potential, because of metal-mercury interactions; for example for the above reaction, the deposition of Pb on mercury, $E = -0.142 \ V$, rather than $-0.126 \ V$. The potentials for plating of metals into mercury are available in the polarographic literature, such as I. M. Kolthoff and J. J. Lingane, *Polarography*, New York: Interscience, 1952.

From the Nernst equation, (9.25),

$$E = 0.80 - 0.059 \log (1/[Ag^+])$$

$$E = 0.80 - 0.059 \log (10^7) = 0.39 \ V$$

As long as the electrode is held at a potential of 0.39 V (or more negative), the silver ion concentration in solution will be decreased to 10^{-7} M (or less).

Copper will start plating at a potential of

$$E = 0.34 - \frac{0.059}{2} \log (1/[Cu^{++}])$$

$$E = 0.34 - \frac{0.059}{2} \log (10^2) = 0.28 \ V$$

If the potential is maintained more positive than 0.28 V, copper will not deposit. Therefore, for a successful separation, the potential should be adjusted so that

$$0.28 \ V < E < 0.39 \ V$$

Depositions at positive potentials may at first appear surprising. However, when an electrode is adjusted to $+0.39$ V, for instance, although it is positive with respect to the reference electrode, it is still the negative electrode (cathode) in the electrolysis circuit (with respect to the auxiliary electrode). Electrodeposition equilibria can also be affected by formation of precipitates and complex ions. These effects can be taken into account by combining the equilibrium constant expressions for these reactions with those for the electrodeposition (see Problem 9.12).

9–5. MISCELLANEOUS EQUILIBRIA

GAS PHASE REACTIONS

The application of equilibrium theory to reactions occurring in the gas phase, such as

$$2CO(g) + O_2(g) \rightleftarrows 2CO_2(g) \tag{9.26}$$

is exactly the same as for that occurring in solution. The equilibrium constant expression for (9.26), in terms of the concentrations of the reacting species, is

$$\frac{[CO_2]^2}{[CO]^2[O_2]} = K \tag{9.27}$$

It is frequently more convenient in studying gas phase reactions to measure and consider the partial pressure of a gas, rather than its concentration, and to give the equilibrium constant in terms of pressures. For an ideal gas, A, the following equation holds

$$P_A V = n_A RT \tag{9.28}$$

where P_A is the partial pressure of A, V is the volume, n_A is the moles of A, T is the temperature, and R is the gas constant (with a value of 0.0821 liter-atmospheres/°K-mole). Rearranging (9.28), remembering that $[A] = n_A/V$, we obtain

$$P_A = [A]RT \tag{9.29}$$

So that the partial pressure of A in a mixture of gases is proportional to its concentration. Thus (9.27) could be written

$$\frac{P_{CO_2}^2}{P_{CO}^2 P_{O_2}} = K_P \tag{9.30}$$

where K_P is the pressure equilibrium constant, with pressures given in atmospheres (atm). K_P is *not* in general equal to K. Substituting the ideal gas equations like (9.29) into (9.30) and comparing with (9.27) shows that for this case

$$K_P = \frac{K}{RT} \tag{9.31}$$

(9.30) can also be written in terms of the moles of each component, n_{CO_2}, n_{CO}, and n_{O_2}, the total number of moles n_T, and the total pressure, P_T, since for the whole mixture

$$P_T V = n_T RT \tag{9.32}$$

where

$$P_T = P_{CO_2} + P_{CO} + P_{O_2} \tag{9.33}$$

$$n_T = n_{CO_2} + n_{CO} + n_{O_2} \tag{9.34}$$

Using (9.28) for each component in (9.30), remembering from (9.32) that $RT/V = P_T/n_T$, we obtain

$$\frac{n_{CO_2}^2 \, n_T}{n_{CO}^2 \, n_{O_2} \, P_T} = K_P \tag{9.35}$$

For the general reaction

$$aA(g) + bB(g) \rightleftarrows cC(g) + dD(g) \tag{9.36}$$

the following equations hold

$$\frac{P_C^c P_D^d}{P_A^a P_B^b} = K_P = K(RT)^{c+d-a-b} \tag{9.37}$$

$$\frac{n_C^c n_D^d}{n_A^a n_B^b}\left(\frac{P_T}{n_T}\right)^{c+d-a-b} = K_P \tag{9.38}$$

$$n_T = n_A + n_B + n_C + n_D \tag{9.39}$$

The solution of equilibrium problems in the gas phase uses the various K_P expressions, material balance expressions for atoms, and an expression for n_T.

Example 9.11. For the reaction

$$N_2O_4(g) \rightleftarrows 2NO_2(g)$$

at 45°C, $K_P = 0.66$. 1.00 mole of $N_2O_4(g)$ is added to an evacuated flask. The final (equilibrium) pressure is found to be 1.10 atm. Calculate the number of moles of NO_2 present at equilibrium.

From (9.38)

$$\frac{n_{NO_2}^2}{n_{N_2O_4}}\frac{P_T}{n_T} = 0.66$$

$$P_T = 1.10 \text{ atm}$$

The material balance for nitrogen is:

$$\text{moles N added} = 2(1.00) = 2.00 \text{ moles}$$

$$2n_{N_2O_4} + n_{NO_2} = 2.00$$

(Alternately, the material balance for O can be written as

$$4n_{N_2O_4} + 2n_{NO_2} = 4.00$$

This is *not* an additional equation, since it can be obtained by multiplying the N-balance equation by 2)

$$n_T = n_{N_2O_4} + n_{NO_2}$$

Letting $x = n_{NO_2}$

$$n_{N_2O_4} = 1.00 - \frac{x}{2}$$

$$n_T = 1.00 + \frac{x}{2}$$

Combining these equations with the K_P expression yields

$$\frac{x^2}{(1 - x/2)(1 + x/2)} = \frac{0.66}{1.10}$$

and solving for x,

$$x = n_{NO_2} = 0.722 \text{ moles}$$

Note that an increase in pressure for the system, for example by adding an inert gas to the flask, will decrease the number of moles of NO_2 at equilibrium as predicted by Le Chatlier's Law.

Example 9.12. One reaction in the contact process for producing sulfuric acid is

$$2SO_2 + O_2 \rightleftarrows 2SO_3$$

At 1000°K, K_P for this reaction is 3.5. If 0.400 mole of SO_2 and 0.800 mole of O_2 are reacted under a constant pressure of 1.00 atm, how many moles of SO_3 are produced?

The system is governed by the following equations:

$$\frac{n_{SO_3}^2}{n_{SO_2}^2 n_{O_2} P_T} \frac{n_T}{} = 3.5$$

$$P_T = 1.00$$

Material balance for S

$$n_{SO_2} + n_{SO_3} = 0.400$$

Material balance for O (0.800 moles added as SO_2 and 1.600 moles added as O_2)

$$2n_{SO_2} + 3n_{SO_3} + 2n_{O_2} = 2.400$$

$$n_T = n_{O_2} + n_{SO_2} + n_{SO_3}$$

Letting $x = n_{SO_3}$ yields

$$n_{SO_2} = 0.400 - x$$

$$n_{O_2} = 0.800 - \frac{x}{2}$$

$$n_T = 1.200 - \frac{x}{2}$$

$$\frac{x^2(1.200 - x/2)}{(0.400 - x)^2(0.800 - x/2)} = 3.5$$

Rearranging the above equation yields

$$x^3 - 2.4x^2 + 2.02x - 0.359 = 0$$

Solving for x (the approximation $x \ll 0.4$ is not valid, so that this equation must be solved numerically, see Chapter 10) yields

$$x = n_{SO_3} = 0.239 \text{ mole}$$

When dealing with real, rather than ideal gases, the pressure is replaced by the *fugacity*, f, which is the effective partial pressure, or the gas phase analogy to the activity.

MEMBRANE EQUILIBRIA

The equilibrium set up across a membrane that allows some ions to pass through it and not others is of interest in biological systems

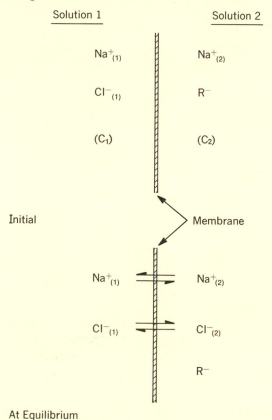

FIGURE 9.4. Membrane equilibrium.

and in the study of some colloids. Consider the initial situation in Figure 9.4, which involves Na^+ and Cl^- in a solution at a concentration C_1 separated by a membrane permeable to Na^+ and Cl^- from a solution containing Na^+ and R^- at a concentration C_2. R^- may be a protein ion which is too large to pass through the membrane, which, in this case, may be the wall of a cell in a biological system. At equilibrium, the sodium ions and chloride ions, which move freely through the membrane, are distributed according to the equation,

$$[Na^+]_{(1)}[Cl^-]_{(1)} = [Na^+]_{(2)}[Cl^-]_{(2)} \tag{9.40}$$

Using this equation as well as electroneutrality and material balance equations, we can predict some of the properties of this situation. The material balances for Na^+ and Cl^- are

$$[Na^+]_{(1)} + [Na^+]_{(2)} = C_1 + C_2 \tag{9.41}$$

$$[Cl^-]_{(1)} + [Cl^-]_{(2)} = C_1 \tag{9.42}$$

The electroneutrality equation in solutions 1 and 2 are

$$[Na^+]_{(1)} = [Cl^-]_{(1)} \tag{9.43}$$

$$[Na^+]_{(2)} = [Cl^-]_{(2)} + [R^-] \tag{9.44}$$

These five equations can be solved rather easily for the five unknowns and yield

$$[Na^+]_{(1)} = \frac{C_1(C_1 + C_2)}{2C_1 + C_2} = [Cl^-]_{(1)} \tag{9.45}$$

$$[Na^+]_{(2)} = \frac{(C_1 + C_2)^2}{2C_1 + C_2} \tag{9.46}$$

$$[Cl^-]_{(2)} = \frac{C_1^2}{2C_1 + C_2} \tag{9.47}$$

$$[R^-] = C_2 \tag{9.48}$$

These results indicate the following. From (9.47), we see that the amount of Cl^- diffusing into 2, $[Cl^-]_{(2)}$, is smaller, the larger the concentration of the nondiffusible ion, C_2. Moreover, comparing (9.45) and (9.46), we see that the total concentration of cation on both sides of the membrane is different. (For example, if $C_1 = C_2 = 0.5\ M$, then at equilibrium, $[Na^+]_{(1)} = 0.33\ M$ and $[Na^+]_{(2)} = 0.67\ M$). As a result of this difference of concentration, an electrical potential exists between solutions 1 and 2 that is called the *membrane potential*. These kinds of equilibria, sometimes called *Donnan equilibria* after their original discoverer, are of importance in biological systems and govern the passage of substances into and out of the cell and the potentials arising in the systems.

OTHERS

Equilibrium theory can be applied to many other situations of interest. For example in the decay of radioactive element through a series of steps involving intermediate radioactive species, a state of equilibrium is reached in which the rate of formation of any element from its parent equals the rate at which it is disintegrating. For this condition equilibrium constant expressions for the intermediates can be written and the amounts of the different radioelements calculated. In the theory of transistors an equilibrium constant expression for the concentration of electrons n, and of holes, p, in the semiconductor can be written as

$$np = N \tag{9.49}$$

where N is a constant. Note that (9.49) is analogous to the water equilibrium constant expression (3.16).

Equilibrium concepts can also be applied to photochemical systems, electrochemical systems, and probably to many new situations yet to be uncovered. In these cases the approach to solving problems will be similar to that presented in the preceding examples.

PROBLEMS

9.1. Consider the extraction of a species A from V_w ml of water (w) with V_o ml of an organic phase (o). Letting $[A]_o/[A]_w = K_D$, $C =$ the initial concentration of A in the aqueous phase before adding organic extractant, and $x =$ millimoles of A left in the aqueous phase after one extraction, derive the following equations:

$$x = \frac{CV_w{}^2}{V_w + K_D V_o} \qquad [A]_w = \frac{CV_w}{V_w + K_D V_o}$$

$$\% \text{ A unextracted} = \frac{100 V_w}{V_w + K_D V_o}$$

9.2. Using the same notation as Problem 9.1, consider the extraction of V_w ml of an aqueous solution of A originally at a concentration C, by repeated extractions with V_o ml of the organic phase. Show that after n extractions, the following equations hold:

$$x_n = CV_w \left[\frac{V_w}{V_w + K_D V_o} \right]^n \qquad [A]_n = C \left[\frac{V_w}{V_w + K_D V_o} \right]^n$$

$$\% \text{ A unextracted} = 100 \left[\frac{V_w}{V_w + K_D V_o} \right]^n$$

9.3. Uranium(VI) can be extracted from aqueous HNO_3 solutions into ethyl ether. When the aqueous phase contains 1.5 M HNO_3 and is saturated in NH_4NO_3, $K_D = 3.5$. Calculate the concentration of U(VI) remaining in the aqueous phase, and the fraction of U(VI) extracted under the following conditions.

(a) 10 ml of 10^{-2} M U(VI) treated with 10 ml of ethyl ether

(b) 10 ml of 10^{-2} M U(VI) treated successively with five 2 ml portions of ethyl ether (see Problem 9.2)

(c) 10 ml of 10^{-3} M U(VI) treated with 100 ml of ethyl ether

9.4. When considering the extraction of a metal chelate, MX_n, from an aqueous phase into an organic phase, the factor D is often used, where

$$D = \frac{\text{Total metal concentration in organic phase}}{\text{Total metal concentration in aqueous phase}}$$

or with the assumption that only MX_n is extracted into the organic phase

$$D = \frac{[MX_n]_o}{[MX_n]_w + [M^{+n}]_w}$$

Use the equilibrium constant expressions, equations (9.3) through (9.6), to derive the following expression

$$D = \frac{K_{D,\ MX_n}}{1 + [H^+]_w^n K_{D,HX}^n / K_a^n K_{stab}^n [HX]_o^n}$$

9.5. For each of the problems below calculate the amount of substance left unreacted in the solution. Assume all resins have capacity of 5 meq/g

(a) 50 ml of 10^{-3} M $CsNO_3$ treated with 1.0 g of 8% crosslinked cation exchange resin in the Na^+-form.

(b) 100 ml of 10^{-3} M $SrCl_2$ treated with 1.0 g of 16% crosslinked cation exchange resin in the H^+-form.

(c) 50 ml of 10^{-3} M KI solution treated with 1.0 g of 8% crosslinked anion exchange resin in the Cl^--form.

(d) 100 ml of 10^{-3} M NaCl solution treated first with 1.0 g of 8% crosslinked cation exchange resin in H^+-form, and then with 1.0 g of 8% crosslinked anion exchange resin in OH^--form.

9.6. An analyst proposes to use a cation exchange technique to determine the total amount of Na^+ and K^+ in a 100 ml solution containing

5×10^{-3} M KCl and 2×10^{-3} M NaCl by treatment with 10 g of an 8% crosslinked cation exchange resin in the H^+-form, with a capacity of 4.0 meq/g and determination of the H^+ liberated. Calculate the concentration of K^+, Na^+, and H^+ in the resulting solution after the treatment.

9.7. Calculate the concentrations of the following ions remaining in solution after 10 ml of a 10^{-3} M solution of each in the indicated concentration of HCl is treated with 1.0 g of Dowex 1-X10 anion exchange resin (see Fig. 9.2 for D-values).

(a) Ti^{++++} in 12 M HCl　　(b) Co^{++} in 8 M HCl

(c) Cd^{++} in 4 M HCl　　(d) Au^{+++} in 2 M HCl

9.8. A method of separating different ions on an ion exchange column called *elution chromatography* involves adding a solution of the ions to an anion exchange resin column in 12 M HCl; some ions are held on the column (have large D-values) and others pass through. Lower concentrations of HCl are then used to *elute* some of the ions from the column and leave others still adsorbed. For example, if a 12 M HCl solution containing Ca^{++}, Fe^{++}, and Cu^{++} is added to the column, Ca^{++} will not adsorb, and will pass through. If the column is then eluted with 5 M HCl, Fe^{++} will be removed, but Cu^{++} will stay adsorbed. Finally, elution with 1 M HCl will remove Cu^{++}. Using the data in Figure 9.2, describe how the following mixtures of ions might be separated using the described technique.

(a) Fe^{+++}, Co^{++}, Ni^{++}　　(b) Ti^{++++}, V^{+5}, Zn^{++}

(c) Ag^+, Rh^{+++}, Ru^{++++}　　(d) Ge(IV), Ga(III), Al(III)

9.9. Calculate the concentration of metal ion left undeposited for the following electrodepositions.

(a) Cu plated from a 10^{-1} M solution of Cu^{++} on a platinum electrode at a potential of 0.22 V.

(b) Zn plated from 100 ml of a 1.00×10^{-2} M Zn^{++} solution using 10.0 ml of mercury as a cathode at a potential of -0.85 V.

(c) Cd plated from 100 ml of a 2.00×10^{-3} M Cd^{++} solution using 20.0 ml of mercury as a cathode at a potential of -0.55 V.

(d) Sn plated from a 10^{-2} M solution of Sn^{++} on a gold cathode at a potential of -0.254 V.

9.10. An analyst wants to separate Cd from Zn in a 100 ml solution containing 0.0100 M Zn^{++} and 1.0×10^{-3} M Cd^{++} by depositing Cd on a mercury cathode with a volume of 10.0 ml. What potential range can be used to separate Cd from Zn quantitatively (i.e., leave less than 0.1% Cd^{++} in solution and deposit less than 0.1 %Zn)?

9.11. A solution contains 10^{-2} M concentrations of Ag^+, Cu^{++}, Pb^{++}, and Cd^{++}. Describe how successive depositions of the metals could be accomplished using a platinum cathode. Indicate the potential to be used for each step.

9.12. Consider the deposition of a metal ion, M^{+n}, on a solid electrode

$$M^{+n} + ne \rightleftarrows M$$

in the presence of a ligand, X^-, which forms a complex ion with M^{+n}

$$M^{+n} + pX^- \rightleftarrows MX_p^{n-p}$$

Show that the equation governing the deposition under these conditions is

$$E = E^0 - \frac{0.06}{n} \log \frac{[MX_p^{n-p}]}{[X^-]^p} - \frac{0.06}{n} \log K_{stab}$$

9.13. The reaction

$$N_2(g) + O_2(g) \rightleftarrows 2NO(g)$$

occurs at high temperatures (e.g., during a lightning flash) and causes the fixation of atmospheric nitrogen. At 2000° K for this reaction, $K_P = 2 \times 10^{-4}$. When this reaction occurs in air, with $P_{N_2} = 0.80$ atm and $P_{O_2} = 0.20$ atm, what is the resulting P_{NO} and [NO] at equilibrium?

9.14. For the reaction for the production of ammonia

$$N_2(g) + 3H_2(g) \rightleftarrows 2NH_3(g)$$

$K_P = 1.7 \times 10^{-4}$ at 673° K. Calculate the number of moles of NH_3 formed when 5.0 moles of N_2 and 3.0 moles of H_2 are reacted at a constant pressure of 1 atm.

9.15. Repeat Problem 9.14 for a constant pressure of 10 atm (a numerical or graphical method of solution must be used, see Chapter 10).

SUPPLEMENTARY READING

LIQUID–LIQUID EXTRACTION

Irving, H., and R. J. P. Williams, in I. M. Kolthoff and P. J. Elving (eds.), *Treatise on Analytical Chemistry*, New York: Interscience, 1961, chap. 31.

Laitinen, H. A., *Chemical Analysis*, New York: McGraw-Hill, 1950, pp. 258–274.

Morrison, G. H., and H. Freiser, *Solvent Extraction in Analytical Chemistry*, New York: Wiley, 1957.

ION EXCHANGE

Rieman, N., and A. C. Breyer, in I. M. Kolthoff and P. J. Elving (eds.), *Treatise on Analytical Chemistry*, New York: Interscience, 1961, chap. 35.

Samuelson, O., *Ion Exchangers in Analytical Chemistry*, New York: Wiley, 1953.

Walton H. F., *Principles and Methods of Chemical Analysis*, Englewood Cliffs, N.J.: Prentice-Hall, 1964, chap. 8.

ELECTRODEPOSITION

Lingane, J. J., *Electroanalytical Chemistry*, New York: Interscience, 1958.

GAS PHASE REACTIONS

Glasstone, S., *Textbook of Physical Chemistry*, New York: Van Nostrand, 1964, pp. 817–842.

CHAPTER 10 NUMERICAL AND COMPUTER METHODS

10–1. INTRODUCTION

Most of the problems considered in the preceding chapters could be solved by making simple approximations. The graphical method of representing equilibrium systems in Chapter 8 allowed treatment of more complicated systems and calculation of approximate answers. Sometimes, however, simple approximations do not hold, or the system is not amenable to graphical representation. In these cases rigorous solution of the equations describing the system is required. The solution of higher degree algebraic equations generally involves graphical or numerical methods, which are discussed in the first part of this chapter. The application of digital computer techniques to the solution of equilibrium problems is then briefly considered. The examples used involve rather simple equations so that the reader can follow the methods without difficulty. The extension of the methods to more complex problems is straightforward.

10–2. ALGEBRAIC SOLUTIONS

QUADRATIC EQUATIONS

The dissociation of a C M solution of an acid HA with an ionization constant K_a is described by the following equations (Chapter 3):

$$[H^+][A^-]/[HA] = K_a \qquad (10.1)$$

$$[H^+][OH^-] = K_w \qquad (10.2)$$

$$[H^+] = [A^-] + [OH^-] \qquad (10.3)$$

$$[HA] + [A^-] = C \qquad (10.4)$$

Under the conditions that the $[OH^-]$ is much smaller than $[H^+]$, the $[OH^-]$-term is dropped from (10.3) and the following equation results:

$$\frac{x^2}{(C-x)} = K_a \qquad (10.5)$$

where x is $[H^+]$. This may be rewritten

$$x^2 + K_a x - K_a C = 0 \qquad (10.6)$$

This is a second degree, or quadratic, equation of the general form

$$ax^2 + bx + c = 0 \qquad (10.7)$$

which can be solved using the quadratic formula

$$x = \frac{-b \pm \sqrt{b^2 - 4ac}}{2a} \qquad (10.8)$$

Example 10.1. Solve (10.6) for $C = 1.00$ M and $K_a = 2.00$. Application of (10.8) yields

$$x = \frac{-2.00 + \sqrt{4.00 + 8.00}}{2} = \frac{-2.000 + 3.464}{2}$$

$$x = 0.732 \quad Answer$$

CUBIC AND HIGHER DEGREE EQUATIONS

Combination of (10.1) through (10.4) with no approximations yields

$$x^3 + K_a x^2 - (K_w + K_a C)x - K_a K_w = 0 \qquad (10.9)$$

This is a third degree, or cubic, equation. Although formulas for solving cubic equations exist,* it is often more convenient to obtain solutions to cubic equations by numerical methods. Problems involving polyprotic acids or several simultaneous equilibria yield fourth or larger degree equations. These can only be solved by numerical methods.

10–3. GRAPHICAL METHODS

In a graphical method of solution, one or more functions are plotted as a function of the variable and the location of the place on the graph where the equation of interest is satisfied is the solution. The graphical methods described here, in contrast to those in Chapter 8, involve the algebraic combination of the equations, and then graphical solution. For example, to solve (10.6) graphically, $x^2 + K_a x - K_a C$ may be calculated for different values of x, and these values plotted on a graph against x. Where the curve crosses the X axis, (10.6) is satisfied, and the value of x at this point is the solution. In general, the solution of an equation

$$F(x) = 0 \qquad\qquad (10.10)$$

where $F(x)$ is some function of x, is obtained by plotting $F(x)$ against x and finding where the curve crosses the X axis. Another method of obtaining a graphical solution to (10.6) is to rearrange it as

$$x^2 = K_a(C - x) \qquad\qquad (10.11)$$

and plot x^2 against x, and $K_a(C - x)$ against x. The intersection of these two curves is the place where (10.11) is satisfied, and the value of x at this location is the solution to the problem. In general, the solution to an equation written in the form

$$f(x) = g(x) \qquad\qquad (10.12)$$

is obtained by plotting $f(x)$ and $g(x)$ against x, and finding the intersection of the two curves.

Example 10.2. Find the solution of the equation in Example 10.1 graphically.

The equation to be solved is

$$x^2 + 2.00x - 2.00 = 0 \qquad\qquad (10.13)$$

A plot of the left-hand side of (10.13) vs. x is shown in Figure 10.1(a) and the solution is $x = 0.73$.

* See, for example, I. S. Sokolnikoff and E. S. Sokolnikoff, *Higher Mathematics for Engineers and Physicists*, New York: McGraw-Hill, 1941, pp. 86–91.

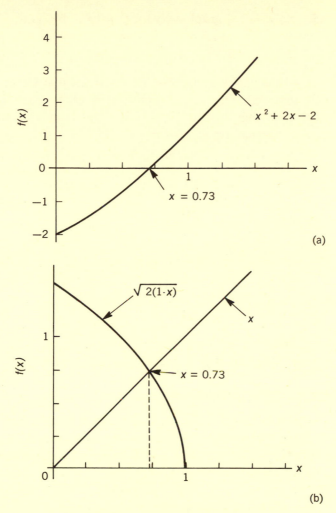

FIGURE 10.1.   Graphical solution to example
10.2 by plotting (a) $x^2 + 2x - 2$ vs. x; (b) $\sqrt{2 - 2x}$
and x vs. x.

This equation can also be solved by rearranging to

$$x^2 = 2 - 2x$$

and plotting x^2 and $2 - 2x$ vs. x. Another possibility is

$$x = \sqrt{2 - 2x}$$

and plotting x and $\sqrt{2 - 2x}$ vs. x (Figure 10.1(b)), and finding the
solution at the intersection of the curves.

Example 10.3. Solve (10.9) graphically for $C = 10^{-6}$ M and $K_a = 10^{-8}$. (10.9) becomes

$$x^3 + 10^{-8}x^2 - (2 \times 10^{-14})x - 10^{-22} = 0 \qquad (10.14)$$

This equation will be easier to handle if the exponential terms in the coefficients are removed. This is accomplished by using a substitution, such as $y = 10^7 x$, so that $x = 10^{-7}y$, $x^2 = 10^{-14}y^2$, and $x^3 = 10^{-21}y^3$. These substitutions transform (10.14) to

$$10y^3 + y^2 - 20y - 1 = 0 \qquad (10.15)$$

This equation can be solved graphically by plotting (10.15) as a function of y and observing where the function crosses the axis. An alternate method involves rearranging (10.15) to, for example,

$$y^2 = \frac{20y + 1}{10y + 1}$$

A plot of y^2 and $(20y + 1)/(10y + 1)$ vs. y yields two curves which intersect at $y = 1.4$, so that

$$x = 10^{-7}y = 1.4 \times 10^{-7} \quad \textit{Answer}$$

10–4. NUMERICAL METHODS

SIMPLE ITERATIVE METHODS

Most numerical methods for solving equations are based on an initial guess for the answer (x_0), followed by some procedure for obtaining a series of other values, $x_1, x_2, \dots$. A suitable method yields values of $x_1, x_2, \dots$ which come closer and closer to the answer; in this case the values are said to *converge* to the required root. If the values get further and further from the answer, the sequence of values is said to *diverge*.

One simple iterative method, sometimes called the method of *successive approximations*, involves writing the equation of interest in the form

$$x = f(x) \qquad (10.16)$$

A value of x_0 is guessed, this value is used in the right-hand side of (10.16) to calculate a new value of x, called x_1, which is used, in turn, to calculate x_2, etc. In this method then, each new x value is calculated from the equation

$$x_{n+1} = f(x_n)$$

Hopefully the sequence of values converge to the answer.

Example 10.4. Solve (10.6) with $K_a = 0.100$ and $C = 1.00$ M by the method of successive approximations.

Rearranging (10.6), we obtain

$$x^1 = \sqrt{0.100(1.00 - x)}$$

which is in the form of (10.16). Remembering that x represents the $[H^+]$ in a 1 M solution of HA, x must lie between 0 and 1. Let us assume $x_0 = 0.100$. Then

$$x_1 = \sqrt{0.100(1.00 - 0.100)} = 0.300$$

$$x_2 = \sqrt{0.100(1.00 - 0.300)} = 0.265$$

$$x_3 = \sqrt{0.100(1.00 - 0.265)} = 0.271$$

$$x_4 = \sqrt{0.100(1.00 - 0.271)} = 0.270 \quad Answer$$

In this case the values rapidly converged to the answer. The reader can verify that any initial guess between 0 and 1 will converge to the answer.

Example 10.5. Solve the equation of Example 10.1 by the method of successive approximations.

Rearranging this equation as before yields

$$x = \sqrt{2.00(1.00 - x)}$$

Choosing $x_0 = 0.50$, we obtain for successive values of x

$$x_1 = \sqrt{2.00(1.00 - 0.50)} = 1.00$$

$$x_2 = \sqrt{2.00(1.00 - 1.00)} = 0.00$$

$$x_3 = \sqrt{2.00(1.00 - 0.00)} = 1.41$$

$$x_4 = \sqrt{2.00(1.00 - 1.41)} = imaginary$$

In this case, the successive values diverge. The reader may verify that any initial guess will lead to divergent values.

The criterion for convergence by this method is that $|f'(x)| < 1$, where $f'(x)$ is the first derivative of the function (or the slope of the function) evaluated at the correct value of x. This is illustrated geometrically in Figure 10.2. For the value x_0, $f(x_0)$ is calculated, leading to x_1, etc. The basis for the rule regarding $f'(x)$ is shown in Figure 10.3. The reader may verify that the successive values of x converge (although at a very leisurely rate) when the equation in Example 10.5 is written in the form

$$x = \frac{(2 - x^2)}{2}$$

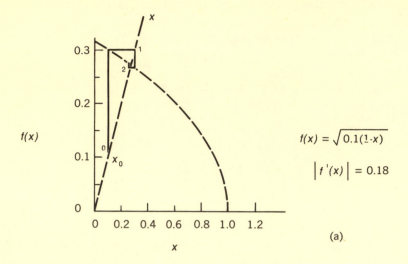

$$f(x) = \sqrt{0.1(1\text{-}x)}$$

$$\left| f'(x) \right| = 0.18$$

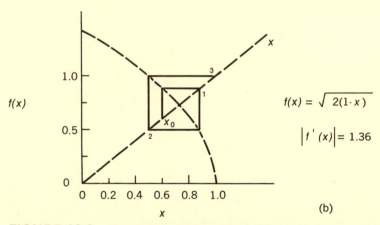

$$f(x) = \sqrt{2(1\cdot x)}$$

$$\left| f'(x) \right| = 1.36$$

FIGURE 10.2. Convergence in the method of successive approximations. (a) convergence in Example 10.4; (b) divergence in Example 10.5; note that curves plotted in this figure are the same as those in Figure 10.1 (b).

WEGSTEIN'S METHOD

A method of successive approximation that is not subject to the limitation of $|f'(x)| < 1$ has been described by Wegstein.* In this method a second sequence of values, $\bar{x}_0$, $\bar{x}_1$, ... , is calculated and used to calculate new values of x. To apply this method, a value of x_0 is chosen.

* J. H. Wegstein, *Comm. Assoc. Comp. Mach.*, June, 1958.

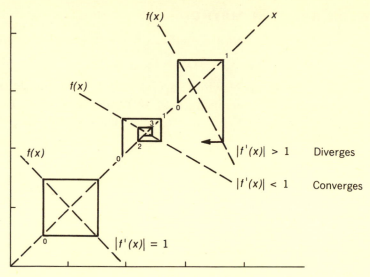

FIGURE 10.3. Geometric basis for criterion for convergence in method of successive approximations.

Then $\bar{x}_0 = x_0$, $x_1 = f(\bar{x}_0)$, $\bar{x}_1 = x_1$, $x_2 = f(\bar{x}_1)$,

$$\bar{x}_2 = x_2 - \frac{(x_2 - x_1)^2}{x_2 - 2x_1 + x_0}$$

This is continued as follows

$$x_n = f(\bar{x}_{n-1})$$

$$\bar{x}_n = x_n - \frac{(x_n - x_{n-1})(x_n - \bar{x}_{n-1})}{x_n - x_{n-1} - \bar{x}_{n-1} + \bar{x}_{n-2}}$$

until $x_n = \bar{x}_{n-1}$. Although this method appears complicated, it is useful in cases where the algebra and calculus necessary for other methods is tedious.

Example 10.6. Solve the equation in Example 10.5 using Wegstein's method.

Applying the above equations, the following successive values are obtained

$$x_0 = 0.600 \quad \bar{x}_0 = 0.600 \quad x_4 = 0.7343 \quad \bar{x}_4 = 0.7341$$

$$x_1 = 0.8944 \quad \bar{x}_1 = 0.8944 \quad x_5 = 0.7292 \quad \bar{x}_5 = \text{p.7269}$$

$$x_2 = 0.4596 \quad \bar{x}_2 = 0.7188 \quad x_6 = 0.7391 \quad \bar{x}_6 = 0.7321$$

$$x_3 = 0.7500 \quad \bar{x}_3 = 0.7304 \quad x_7 = 0.7320 \quad \textit{Answer}$$

NEWTON-RAPHSON METHOD

Another group of iterative methods makes use of the formula

$$x_{n+1} = x_n - F(x_n)/m \qquad (10.17)$$

instead of (10.16). In the Newton-Raphson method m is the slope of the curve $F(x)$ at x_n. The significance of this method is illustrated geometrically in Figure 10.4. A value of x_0 is guessed. The slope of the

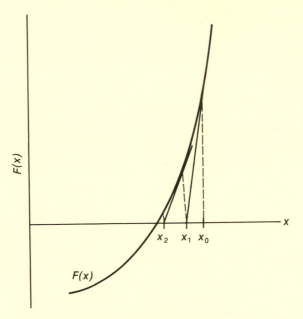

FIGURE 10.4. Geometric representation of Newton-Raphson method.

curve at x_0, which is $F'(x_0)$, is calculated. Since the slope is given by

$$m = F'(x_0) = \frac{F(x_0) - 0}{x_0 - x_1} \qquad (10.18)$$

x_1 can be calculated by rearranging (10.18) to

$$x_1 = x_0 - \frac{F(x_0)}{F'(x_0)} \qquad (10.19)$$

and x_1 lies closer to the root of $F(x)$ than x_0. This process is continued until the root is found. The criterion for convergence for this method is $F'(x) \neq 0$ and $F''(x) \neq 0$ at the root of $F(x)$; these are less restrictive conditions than that of the successive approximations method. The

Newton-Raphson method also leads to a more rapid convergence than the simple iterative methods.

Example 10.7. Solve the equation in Example 10.5 by the Newton-Raphson method.

$$F(x) = x^2 + 2x - 2 \quad F'(x) = 2x + 2$$

$$x_0 = 0.600 \quad F(x_0) = -0.440 \quad F'(x_0) = 3.200$$

$$x_1 = 0.600 - (-0.440)/(3.200) = 0.7375$$

$$F(x_1) = 0.0189 \quad F'(x_1) = 3.475$$

$$x_2 = 0.7375 - (0.0189)/(3.475) = 0.7321$$

This is the answer, since $F(x_2) = 0.00017$. The reader may try other values for x_0, for example 0 or 1, and show that these also rapidly converge to the root.

The Newton-Raphson method can also be applied to simultaneous equations in two unknowns by a modification of the described procedure. Another iterative method based on (10.17), called the method of false position, requires two guesses for x, between which the required root lies. In this method m is the slope of the line connecting these points. This method, and many others, are described in the references on numerical methods given in the bibliography at the end of this chapter.

10–3. COMPUTER METHODS

The numerical and graphical methods described are useful for solving simpler equilibrium problems. However, as problems become more complex, involving more species in the equilibrium mixture and a larger number of equilibrium constant expressions, combining the algebraic equations becomes tedious and making reasonable approximations becomes more difficult. For complicated equilibria then, one turns to techniques and programs suitable for computations on a desk calculator, or preferably, on a high speed digital computer. Interest in these methods has been stimulated by the importance of equilibrium calculations in analytical problems and particularly in problems involving the combustion of fuels and propellants. While it is beyond the scope of this treatment to give a detailed discussion of the different techniques proposed, a brief outline of some recent techniques will be given.

FORMULATION OF A COMPUTER METHOD

Equilibrium problems involve the solution of a set of equations, some of which are nonlinear. Although rather straightforward methods, based on matrix methods, exist for solving sets of simultaneous linear equations, nonlinear simultaneous equations must be solved by numerical methods. Furthermore, since a computer is incapable of operating in terms of unknown variables and can only perform arithmetic operations with numbers, it cannot do algebraic manipulations. Therefore the general approach used in the solution of equilibrium problems on a computer is based on the following steps.

1. Guesses of some of the concentrations are made.
2. The equilibrium constant and material balance equations are used to calculate the other unknown concentrations, using the guessed values and previously calculated concentrations.
3. A test is made to see if all of the equations are obeyed.
4. The values of the concentrations are varied and the process repeated until all equations are satisfied to the desired accuracy.

The different methods used generally follow the procedure given above, but differ in their choice and in the formulation of the equations, the initial guess method, and the iterative procedure used. Although most of the proposed computer methods have been concerned with gas phase equilibria and use the formulations given in 9–5, the extension of these methods to ionic equilibria should prove to be straightforward.

BRINKLEY'S METHOD

The method proposed by Brinkley and discussed by Kandiner and Brinkley is based upon writing the usual equilibrium constant expressions and material balance equations. Some of the chemical species, usually equal in number to the number of different chemical elements involved in the problem, are chosen as the "independent components" (also called the "basis"). These are generally the components having the greatest probable concentrations at equilibrium. The equations are then used to calculate the concentrations of the other species (the "derived constituents"). Initial values for the independent components are usually calculated from the material balance equations by assuming that the derived constituents are absent. These values are improved using the method of successive approximations or the Newton–Raphson method, until the difference between successive values of each independent component is less than some arbitrary tolerance. Tabular methods are given for writing equations and the extension to problems involving nonideality and multiphase systems is discussed. Variations of this method have been proposed by Huff and coworkers and by Browne.

VILLARS' METHOD

In this method, equilibrium constant expressions are written for derived constituents in terms of certain independent components. The equilibrium constant expressions are treated one at a time, neglecting the interactions with the other equilibria. Changes in composition are calculated for the reaction system showing the greatest discrepancy in its equilibrium relationship and the amounts of the different species are revised on the basis of the new results. The process is repeated until the maximum discrepancy is less than a preassigned error. The method has been re-evaluated by Cruise, who modified it to make the optimum choice of independent components and to speed convergence.

WHITE'S METHOD

Equations are written in terms of the free energies in this method. Since the total free energy of a system is smallest at equilibrium (Section 1–3), the equilibrium amounts of the constituents can be calculated by finding those amounts which minimize the total free energy subject to the constraint of the material balance equations. The methods of steepest descent and linear programming are used to minimize the free energy function.

THE METHOD OF ANTHONY AND HIMMELBLAU

The set of equations

$$f_j(x_1, x_2, \ldots x_m) = 0$$

comprising the equilibrium constant expressions and the material balance equations are solved using the "Direct Search" technique. In this method the equations are squared and summed to yield a function Φ,

$$\Phi = \Sigma_j f_j(x_1, x_2, \ldots x_m)^2$$

The problem is solved by determining the concentrations (i.e., x values) which minimize Φ. The method operates by choosing an initial set of x values and determing Φ. Each x value is now changed by a small amount, Δx, in an "exploratory move" and Φ is checked to ascertain whether the Δx has made Φ smaller. On each trial, if Φ is lower, the new value of x is adopted; if not, the sign on Δx is changed and the trial repeated. The process is continued using "exploratory moves" and "pattern moves" until Φ is minimized. Anthony and Himmelblau also point out that the relative weighting of each of the equations in a numerical scheme can affect the answers. For example they repeated the calculations of Kandiner and Brinkley and obtained quite different results, especially for some of the minor species, by varying the relative weights of the different equations in the Φ function.

THE METHOD OF BARD AND KING

This method is an attempt to write a general program for solving ionic equilibria problems using a trial-and-error procedure. Although methods based on trial-and-error procedures are generally slower than those using other numerical methods, the problems of convergence and initial guesses are greatly decreased. This method involves writing the equilibrium constant equations, the material balance equations, and the electro neutrality equation in a form closely resembling that used in the previous chapters. Two species, called the "master variables," that appear most often in the equations, are chosen. The equations are written in a form that can be used to solve for the concentrations of the other species in terms of the master variables and species already calculated. For example, consider the problem of a C M solution of a weak acid H_3A. The six species in this system are H^+, OH^-, H_3A, H_2A^-, HA^{--} and A^{---}, and the equations are

$$K_1 = [H^+][H_2A^-]/[H_3A] \qquad K_2 = [H^+][HA^{--}]/[H_2A^-]$$

$$K_3 = [H^+][A^{---}]/[HA^{--}] \quad K_w = [H^+][OH^-]$$

$$C = [H_3A] + [H_2A^-] + [HA^{--}] + [A^{---}]$$

$$[H^+] = [H_2A^-] + 2[HA^{--}] + 3[A^{---}] + [OH^-]$$

If $[H^+]$ and $[A^{---}]$ are chosen as the master variables, then the equations can be arranged in the following sequence to solve for the concentrations of the other four species:

$$[H_3A] = [H^+]^3[A^{---}]/K_1K_2K_3$$

$$[HA^{--}] = [H^+][A^{---}]/[H_3A]$$

$$[H_2A^-] = K_1[H_3A]/[H^+]$$

$$[OH^-] = K_w/[H^+]$$

The other two equations are used as test equations

$$Y = [H_3A] + [H_2A^-] + [HA^{--}] + [A^{---}] - C$$
$$Y = [H_2A^-] + 2[HA^{--}] + 3[A^{---}] + [OH^-] - [H^+]$$

When the correct concentrations of all six species are obtained, the Y values in the test equations are zero. Initial values of the master variables are guessed as values known to be smaller than the probable values. These are then systematically varied in steps of 10, 1, 0.1, 0.01, ... , until all equations are obeyed. Although this seems to be a tedious and slow procedure, the above problem for 0.2 M H_3PO_4 was solved in only 17 seconds on a high speed digital computer (a Control Data Corporation 1604 computer). A more complicated problem concerning Ni^{++} in an ammonia solution containing EDTA (H_4Y), which involved 17 solution species, was solved in 80 seconds for four different concentrations of EDTA.

Because computers can be programmed to generate data and are also capable of plotting results directly and rapidly, they will probably be much used in the future for performing equilibrium calculations.

PROBLEMS

10.1. Solve the following equations for the real, positive, root to two significant figures using a graphical method.

(a) $[H^+]^2 + (10^{-2})[H^+] - 10^{-2} = 0$

(b) $[Ag^+]^3 + 0.020\,[Ag^+]^2 = 3.0 \times 10^{-5}$

(c) $4[Mg^{++}]^3 + (4.0 \times 10^{-3})\,[Mg^{++}]^2 + (1.0 \times 10^{-6})[Mg^{++}] - (6.0 \times 10^{-9}) = 0$

(d) $\dfrac{(2-y)y^2}{(1-2y)^2(1-y)} = 3.5$

(e) $[OH^-]^4 + (10^{-5})\,[OH^-]^3 + (10^{-11})\,[OH^-]^2 + (10^{-14})\,[OH^-] - (10^{-20}) = 0$

10.2. Solve the equations in Problem 10.1 for the real, positive root to three significant figures using a numerical method.

10.3. Choose the master variables and set up the sequence of equations which can be used in the Bard-King digital computer program method for the following problems.

(a) 0.10 mole of H_4Y (EDTA) diluted to 1.00 liter

(b) 0.10 mole of AgCl treated with 1.0 liter of 0.10 M NH_3

(c) 0.10 mole of H_3PO_4 and 0.15 mole of NH_3 diluted to 1.0 liter

SUPPLEMENTARY READING

NUMERICAL METHODS

Buckingham, R. A., *Numerical Methods*, London: Pitman, 1962.

Eberhart, J. G., and T. R. Sweet, The Numerical Solution of Equations in Chemistry, *J. Chem. Educ.*, **37**, 422 (1960).

Hamming, R. W., *Numerical Methods For Scientists and Engineers*, New York: McGraw-Hill, 1962.

Scarborough, J. B., *Numerical Mathematical Analysis*, Baltimore: The Johns Hopkins Press, 1958.

COMPUTER METHODS

The following is a list of some recent papers referred to in the text. References to other works in this field may be found in these.

Anthony, R. G., and D. M. Himmelblau, Calculation of Complex Chemical Equilibria by Search Techniques, *J. Phys. Chem.*, **67**, 1080 (1963).

Bard, A. J., and D. M. King, General Digital Computer Program For Chemical Equilibrium Calculations, *J. Chem. Educ.*, **42**, 127 (1965).

Brinkley, S. R., The Equilibrium Composition of Systems of Many Constituents, *J. Chem. Phys.*, **14**, 563, 686 (1946); **15**, 107 (1947).

Browne, H., "The Theoretical Computation of Equilibrium Compositions, Thermodynamic Properties, and Performance Characteristics of Propellant Systems," U.S. Naval Ordnance Test Station, NOTS TP 2434 (1960).

Cruise, D. R., "Notes on the Rapid Computation of Chemical Equilibria," *J. Phys. Chem.*, **68**, 3797 (1964).

Huff, V. N., Gordon, S., and V. E. Morrell, "General Method of Thermodynamic Tables for Computation of Equilibrium Composition and Temperature of Chemical Reactions," NACA Report 1037 (1951).

Kandiner, H. J., and S. R. Brinkley, Calculation of Complex Equilibrium Relations, *Ind. Eng. Chem.*, **42**, 850 (1950).

Villars, D. S., A Method of Successive Approximations For Computing Combustion Equilibria on a High Speed Digital Computer, *J. Phys. Chem.*, **63**, 521 (1959).

White, W. B., Johnson S. M., and G. B. Dantzig, Chemical Equilibrium in Complex Mixtures, *J. Chem. Phys.*, **28**, 751 (1958).

APPENDIX A CALCULATION OF CONCENTRATIONS AND REACTION VELOCITIES DURING THE H_2—I_2 REACTION

For the reaction

$$H_2 + I_2 \rightleftarrows 2HI \qquad (1)$$

v_f is the rate at which HI is produced or the rate at which I_2 and H_2 are consumed. Therefore v_f is the rate of change of $[H_2]$ and $[I_2]$ with time, t, or, using calculus

$$v_f = \frac{-d[H_2]_t}{dt} = k_f[H_2]_t[I_2]_t \qquad (2)$$

In our discussion, the initial concentrations of H_2 and I_2 are equal, so that

$$[H_2]_{t=0} = [I_2]_{t=0} = C_i \qquad (3)$$

Since one mole of I_2 is consumed per mole of H_2 consumed, at all times

$$[H_2]_t = [I_2]_t \qquad (4)$$

and (2) may be rewritten

$$\frac{-d[H_2]_t}{dt} = k_f[H_2]_t^2 \tag{5}$$

We can solve this differential equation, using the initial condition

$$[H_2]_{t=0} = C_i \quad \text{at} \quad t=0 \tag{6}$$

$$\int_{C_i}^{[H_2]_t} \frac{d[H_2]_t}{[H_2]_t^2} = -k_f \int_0^t dt \tag{7}$$

$$\frac{1}{[H_2]_t} - \frac{1}{C_i} = k_f t \tag{8}$$

$$[H_2]_t = [I_2]_t = \frac{1}{k_f t + 1/C_i} \tag{9}$$

For $k_f = 2.3$ liter/mole-second, $k_b = 0.14$ liter/mole-second, and $C_i = 10^{-3}$ mole/liter, we can derive the following expressions:

$$[H_2]_t = [I_2]_t = \frac{1}{2.3t + 10^3} \text{ mole/liter} \tag{10}$$

$$v_f = k_f[H_2]_t[I_2]_t = \left[\frac{1}{2.3t + 10^3}\right]^2 2.3 \text{ mole/liter-sec} \tag{11}$$

$$[HI]_t = 2(C_i - [H_2]_t) = 2(10^{-3} - [H_2]_t) \text{ mole/liter} \tag{12}$$

$$v_b = k_b[HI]_t^2 = 0.14(4)(10^{-3} - [H_2]_t)^2 \tag{13}$$

$$v_{net} = v_f - v_b \tag{14}$$

The values for the various quantities $[H_2]_t$, $[I_2]_t$, $[HI]_t$, v_f, v_b, and v_{net}, shown in Figure 1.1 (a) and (b) were calculated using (10) through (14). Some values of these quantities at various times are shown in Table A.1.

TABLE A. 1.   Concentrations and Reaction Velocities During the H_2—I_2 Reaction

t (sec)	$[H_2]_t$ $[I_2]_t$ mmol/l	v_f mmol/l/sec	$[HI]_t$ mmol/l	v_b mmol/l/sec	v_{net} mmol/l/sec
0	1.000	2.3×10^{-6}	0	0	2.3×10^{-6}
200	0.685	1.08	0.630	0.06×10^{-6}	1.02
400	0.521	0.62	0.958	0.13	0.49
600	0.420	0.41	1.160	0.19	0.22
800	0.352	0.29	1.296	0.24	0.05
870	0.333	0.255	1.334	0.249	0.006

Similarly, when only HI is initially present,

$$\frac{-d[HI]_t}{dt} = k_b[HI]_t^2 \tag{15}$$

Solving (15) with the initial condition

$$[HI]_{t=0} = 2 \times 10^{-3}\ M \quad \text{at} \quad t = 0$$

using the same technique as before, we obtain

$$[HI]_t = \frac{1}{k_b t + 1/[HI]_{t=0}} = \frac{1}{0.14t + 500} \text{ moles/liter} \tag{16}$$

$$v_b = 0.14[HI]_t^2 = 0.14\left[\frac{1}{0.14t + 500}\right]^2 \tag{17}$$

$$[H_2]_t = [I_2]_t = \frac{(2 \times 10^{-3}) - [HI]_t}{2} \tag{18}$$

$$v_f = 2.3[H_2]_t[I_2]_t = 2.3[H_2]_t^2 \tag{19}$$

(16) through (19) were used in plotting Figures 1.1 (c) and 1.1 (d)

APPENDIX B CALCULATION OF EQUILIBRIUM CONSTANTS FROM FREE ENERGY DATA

The free energy of a reaction is the maximum amount of useful work (e.g., electrical or mechanical work) obtained by carrying out the reaction. By defining certain standard free energies of elements or ions and measuring the amount of heat liberated during chemical reactions, the standard free energy of a substance can be determined. This *standard free energy* of a substance, G^0, is the free energy of the substance when present in its *standard state*, usually the state of the substance at one atmosphere pressure for pure liquids, solids, or gases, or for a solute at unit activity (or approximately, at unit molarity). Generally the most stable form of an element in the standard state at 25° C is assigned a free energy of zero. The free energy of a substance A in a mixture at other than standard conditions is given by the equation

$$G_A = G_A^0 R + T \ln [A] \tag{1}$$

where G_A^0 is the standard free energy of pure A, R is a constant, T is the temperature, ln is the natural logarithm (equal to 2.3 $\log_{10}$), and

[A] is the concentration (more precisely, the activity) of substance A in the mixture.

Consider the reaction

$$A_2 + B_2 \rightleftarrows 2AB \tag{2}$$

at 25° C, and assume that A_2 and B_2 are in their standard states (such as gaseous H_2, N_2, or O_2, or solid I_2) so that

$$G^0_{A_2} = G^0_{B_2} = 0 \tag{3}$$

and let us assume a free energy of formation of AB, G^0_{AB}, of -1.00 kcal/mole. This number indicates that experiments have shown that upon reaction of A_2 and B_2 to form AB, a net free energy change of 1.00 kilocalorie occurs per mole of AB formed. The minus sign denotes a liberation of free energy, or that the reaction of A_2 and B_2 to form AB at standard conditions is spontaneous. Values of G^0 for many substances in different states are compiled in tables and handbooks.*

For 1 mole of A_2 and 1 mole of B_2 in a 1 liter flask at 25° C,

$$G_{A_2} = G^0_{A_2} + 1.364 \log [A_2] \tag{4}$$

$$G_{B_2} = G^0_{B_2} + 1.364 \log [B_2] \tag{5}$$

($RT \ln x = 1.364 \log x$ at 25° C). Since $G^0_{A_2} = G^0_{B_2} = 0$, and $[A_2] = [B_2] = 1 \, M$,

$$G = (1)(G_{A_2}) + (1)(G_{B_2}) = 0 \tag{6}$$

After 0.1 mole of A_2 and B_2 have reacted to form 0.2 mole of AB

$$G'_{A_2} = 0 + 1.364 \log (0.9) = G'_{B_2} \tag{7}$$

$$G'_{A_2} = G'_{B_2} = -0.0625 \text{ kcal} \tag{8}$$

$$G'_{AB} = G^0_{AB} + 1.364 \log [AB] \tag{9}$$

$$G'_{AB} = -1.00 + 1.364 \log (0.2) = -1.953 \text{ kcal}$$

$$G' = (0.9)(-0.0625) + (0.9)(-0.0625) \tag{10}$$

$$+ (0.2)(-1.953) = -0.503 \text{ kcal} \tag{11}$$

Calculations at other compositions in this reaction mixture are performed in a similar manner.

To find the relation between the equilibrium constant and the free energy of the reaction without doing point-by-point calculations, we can write a general equation describing the curve. If x moles of A_2 and B_2

---

* An especially useful compilation may be found in W. M. Latimer, *Oxidation Potentials*, Englewood Cliffs, N.J.: Prentice-Hall, 1952.

react to form $2x$ moles of AB, then $n_{A_2} = 1 - x$, $n_{B_2} = 1 - x$, and $n_{AB} = 2x$. The free energy of the system is then

$$G = (1 - x)(G_{A_2}^0 + RT \ln [A_2]) + (1 - x)(G_{B_2}^0 + RT \ln [B_2]) \\ + (2x)(G_{AB}^0 + RT \ln [AB]) \tag{12}$$

Equilibrium is attained in the system when the total free energy of the system, G, is at a minimum. At what value of x will G be smallest? We remember from calculus that to find the condition at which G is at a minimum, for G varying with x, we set

$$\frac{dG}{dx} = 0 \tag{13}$$

or differentiating (12)

$$0 = -(G_{A_2}^0 + RT \ln [A_2]_{eq}) - (G_{B_2}^0 + RT \ln [B_2]_{eq}) \\ + 2(G_{AB}^0 + RT \ln [AB]_{eq}) \tag{14}$$

Rearrangement of (14) yields

$$2G_{AB}^0 - G_{A_2}^0 - G_{B_2}^0 = RT \ln [A_2]_{eq} + RT \ln [B_2]_{eq} - 2RT \ln [AB]_{eq} \tag{15}$$

$$\Delta G^0 = -RT \ln \frac{[AB]_{eq}^2}{[A_2]_{eq}[B_2]_{eq}} = -RT \ln K \tag{16}$$

where

$$\Delta G^0 = 2G_{AB}^0 - G_{A_2}^0 - G_{B_2}^0 \tag{17}$$

$$K = \frac{[AB]_{eq}^2}{[A_2]_{eq}[B_2]_{eq}} \tag{18}$$

For the reaction under consideration,

$$G^0 = 2(-1.00) - 0 - 0 = -2.00 \text{ kcal/mole}$$

$$RT \ln K = 1.364 \log K = -(-2.00)$$

$$K = 29.3$$

For the general reaction

$$aA + bB + \cdots \rightleftarrows cC + dD + \cdots \tag{19}$$

$$\Delta G^0 = -RT \ln K \tag{20}$$

$$\Delta G^0 = cG_C^0 + dG_D^0 + \cdots - aG_A^0 - bG_B^0 - \cdots \tag{21}$$

(20) and (21) allow the calculation of the equilibrium constant of any reaction of compounds with known (i.e., measured) standard free energies. Because free energy data can be collected by methods not

related to the chemical reaction of interest, we can calculate equilibrium constants of reactions that might be difficult or impossible to carry out in a finite time. A more exact (but equivalent) derivation of (20) would use activities rather than concentrations.

APPENDIX C TABLES OF
EQUILIBRIUM CONSTANTS

Data in these tables were obtained from the following more complete compilations:

J. Bjerrum, G. Schwarzenbach, and L. G. Sillen, *Stability Constants of Metal-Ion Complexes, with Solubility Products of Inorganic Substances. Part I: Organic Ligands. Part II: Inorganic Ligands*, Chem. Soc. (London) Spec. Publ. No. 6 (1957); No. 7 (1958), and *Dissociation Constants of Organic Acids in Aqueous Solution*, Pure and Applied Chem., Vol. 1, Nos. 2–3.

The values selected were those determined at or near 25° C and, where possible, were those extrapolated to infinite dilution ($\mu = 0$) so as to represent thermodynamic equilibrium constants. Often several different values are listed in the literature for the equilibrium constant of a given reaction. For example, listed values for the second dissociation constant of H_2S, K_2, differ from one another by several orders of magnitude. Like other experimental measurements, the reliability of equilibrium constant data depends on both the accuracy of the measurement and on the interpretation of the data. The value listed in these tables usually represents the most modern value, although sometimes the choice between several values was quite arbitrary.*

TABLE C. 1.   Dissociation Constants of Some Acids at 25° C (To determine K_b for conjugate base, $K_b = K_w/K_a$)

Acid	Reaction	pK_a	K_a
Inorganic Acids			
Ammonium	$NH_4^+ \rightleftarrows NH_3 + H^+$	9.26	5.5×10^{-10}
Arsenic (1)	$H_3AsO_4 \rightleftarrows H^+ + H_2AsO_4^-$	2.22	6.0×10^{-3}
(2)	$H_2AsO_4^- \rightleftarrows H^+ + HAsO_4^{--}$	6.98	1.0×10^{-7}
(3)	$HAsO_4^{--} \rightleftarrows H^+ + AsO_4^{---}$	11.53	3.0×10^{-12}
Arsenious	$HAsO_2 \rightleftarrows H^+ + AsO_2^-$	9.22	6.0×10^{-10}
Boric	$HBO_2 \rightleftarrows H^+ + BO_2^-$	9.23	5.9×10^{-10}
Carbonic (1)	$H_2CO_3 \rightleftarrows H^+ + HCO_3^-$	6.35	4.5×10^{-7}
(2)	$HCO_3^- \rightleftarrows H^+ + CO_3^{--}$	10.33	4.7×10^{-11}
Chromic (2)	$HCrO_4^- \rightleftarrows H^+ + CrO_4^{--}$	6.50	3.2×10^{-7}
Cyanic	$HCNO \rightleftarrows H^+ + CNO^-$	3.66	2.2×10^{-4}
Hydrazinium	$N_2H_5^+ \rightleftarrows H^+ + N_2H_4$	7.99	1.0×10^{-8}
Hydrocyanic	$HCN \rightleftarrows H^+ + CN^-$	9.14	7.2×10^{-10}
Hydrofluoric	$HF \rightleftarrows H^+ + F^-$	3.17	6.7×10^{-4}
Hydroselenic (1)	$H_2Se \rightleftarrows H^+ + HSe^-$	3.89	1.3×10^{-4}
(2)	$HSe^- \rightleftarrows H^+ + Se^{--}$	11.00	1.0×10^{-11}
Hydrosulfuric (1)	$H_2S \rightleftarrows H^+ + HS^-$	6.96	1.1×10^{-7}
(2)	$HS^- \rightleftarrows H^+ + S^{--}$	14.0	1×10^{-14}
Hydroxylammonium	$NH_3OH^+ \rightleftarrows NH_2OH + H^+$	5.98	1.0×10^{-6}
Hypobromous	$HBrO \rightleftarrows H^+ + BrO^-$	8.68	2.1×10^{-9}
Hypochlorous	$HClO \rightleftarrows H^+ + ClO^-$	7.53	3.0×10^{-8}
Nitrous	$HNO_2 \rightleftarrows H^+ + NO_2^-$	3.29	5.1×10^{-4}
Phosphoric (1)	$H_3PO_4 \rightleftarrows H^+ + H_2PO_4^-$	2.23	5.9×10^{-3}
(2)	$H_2PO_4^- \rightleftarrows H^+ + HPO_4^{--}$	7.21	6.2×10^{-8}
(3)	$HPO_4^{--} \rightleftarrows H^+ + PO_4^{---}$	12.32	4.8×10^{-13}
Sulfuric (2)	$HSO_4^- \rightleftarrows H^+ + SO_4^{--}$	1.99	1.0×10^{-2}
Sulfurous (1)	$H_2SO_3 \rightleftarrows H^+ + HSO_3^-$	1.76	1.7×10^{-2}
(2)	$HSO_3^- \rightleftarrows H^+ + SO_3^{--}$	7.21	6.2×10^{-8}
Organic Acids			
Acetic	$CH_3COOH \rightleftarrows H^+ + CH_3COO^-$	4.74	1.8×10^{-5}
Anilinium	$C_6H_5NH_3^+ \rightleftarrows H^+ + C_6H_5NH_2$	4.61	2.5×10^{-5}
Benzoic	$C_6H_5COOH \rightleftarrows H^+ + C_6H_5COO^-$	4.20	6.3×10^{-5}
Citric	$HOOCCH_2 \cdot C(OH)(COOH) \cdot CH_2COOH = H_3Cit$		
(1)	$H_3Cit \rightleftarrows H^+ + H_2Cit^-$	3.13	7.4×10^{-4}
(2)	$H_2Cit^- \rightleftarrows H^+ + HCit^{--}$	4.76	1.8×10^{-5}
(3)	$HCit^{--} \rightleftarrows H^+ + Cit^{---}$	6.40	4.0×10^{-7}

* See D. DeFord, The Reliability of Calculations Based on the Law of Chemical Equilibrium, *J. Chem. Educ.*, **31**, 460 (1954).

TABLE C. 1. (*Continued*)

Acid	Reaction	pK_a	K_a
	Organic Acids (*continued*)		
Ethylammonium	$C_2H_5NH_3^+ \rightleftharpoons H^+ + C_2H_5NH_2$	10.67	2.1×10^{-11}
Ethylenediammonium(1)	$C_2H_4(NH_3)_2^{++} \rightleftharpoons H^+ +$		
	$\quad C_2H_4NH_3NH_2^+$	7.52	3.0×10^{-8}
(2)	$C_2H_4NH_3NH_2^+ \rightleftharpoons H^+ +$		
	$\quad C_2H_4(NH_2)_2$	10.65	2.2×10^{-11}
Ethylenediaminetetra-	$C_2H_4[N(CH_2COOH)_2]_4 = H_4Y$		
acetic (EDTA) (1)	$H_4Y \rightleftharpoons H^+ + H_3Y^-$	1.99	1.0×10^{-2}
(2)	$H_3Y^- \rightleftharpoons H^+ + H_2Y^{--}$	2.67	2.1×10^{-3}
(3)	$H_2Y^{--} \rightleftharpoons H^+ + HY^{---}$	6.16	6.9×10^{-7}
(4)	$HY^{---} \rightleftharpoons H^+ + Y^{-4}$	10.22	6.0×10^{-11}
Formic	$HCOOH \rightleftharpoons H^+ + HCOO^-$	3.77	1.7×10^{-4}
Methylammonium	$CH_3NH_3^+ \rightleftharpoons CH_2NH_2 + H^+$	10.72	1.9×10^{-11}
Oxalic (1)	$(COOH)_2 \rightleftharpoons H^+ + COOHCOO^-$	1.25	5.6×10^{-2}
(2)	$COOHCOO^- \rightleftharpoons H^+ +$		
	$\quad (COO)_2^{--}$	4.28	5.2×10^{-5}
Phenol	$C_6H_5OH \rightleftharpoons H^+ + C_6H_5O^-$	9.95	1.1×10^{-10}
o-Phthalic (1)	$C_6H_5(COOH)_2 \rightleftharpoons H^+ +$		
	$\quad C_6H_5COOH(COO)^-$	2.95	1.1×10^{-3}
(2)	$C_6H_5COOH(COO)^- \rightleftharpoons H^+ +$		
	$\quad C_6H_5(COO)_2^{---}$	5.41	3.9×10^{-6}
Pyridinium	$C_5H_5NH^+ \rightleftharpoons C_5H_5N + H^+$	5.17	6.8×10^{-6}
Tartaric	$(CHOH \cdot COOH)_2 = H_2Tar$		
(1)	$H_2Tar \rightleftharpoons H^+ + HTar^-$	3.04	9.1×10^{-4}
(2)	$HTar^- \rightleftharpoons H^+ + Tar^{--}$	4.37	4.3×10^{-5}

TABLE C. 2.   Solubility Product Constants of Some Compounds at 25° C

Compound	Reaction	pK_{sp}	K_{sp}
Aluminium			
Hydroxide	$Al(OH)_3 \rightleftharpoons Al^{+++} + 3OH^-$	32	1×10^{-32}
Phosphate	$AlPO_4 \rightleftharpoons Al^{+++} + PO_4^{---}$	18.2	6.3×10^{-19}
Barium			
Arsenate	$Ba_3(AsO_4)_2 \rightleftharpoons 3Ba^{++} + 2AsO_4^{---}$	50.1	8×10^{-51}
Carbonate	$BaCO_3 \rightleftharpoons Ba^{++} + CO_3^{--}$	8.3	5×10^{-9}
Chromate	$BaCrO_4 \rightleftharpoons Ba^{++} + CrO_4^{--}$	9.93	1.2×10^{-10}
Fluoride	$BaF_2 \rightleftharpoons Ba^{++} + 2F^-$	5.98	1.05×10^{-6}
Iodate	$Ba(IO_3)_2 \rightleftharpoons Ba^{++} + 2IO_3^-$	8.82	1.5×10^{-9}
Manganate	$Ba(MnO_4)_2 \rightleftharpoons Ba^{++} + 2MnO_4^-$	9.61	2.5×10^{-10}
Oxalate	$BaC_2O_4 \rightleftharpoons Ba^{++} + C_2O_4^{--}$	7.82	1.5×10^{-8}
Sulfate	$BaSO_4 \rightleftharpoons Ba^{++} + SO_4^{--}$	10.00	1.0×10^{-10}

TABLE C. 2. (*Continued*)

Compound	Reaction	pK_{sp}	K_{sp}
Bismuth			
Arsenate	$BiAsO_4 \rightleftarrows Bi^{+++} + AsO_4^{---}$	9.4	4×10^{-10}
Hydroxide	$Bi(OH)_3 \rightleftarrows Bi^{+++} + 3OH^-$	30.4	4×10^{-31}
Phosphate	$BiPO_4 \rightleftarrows Bi^{+++} + PO_4^{---}$	22.9	1.2×10^{-23}
Sulfide	$Bi_2S_3 \rightleftarrows 2Bi^{+++} + 3S^{--}$	97	1×10^{-97}
Cadmium			
Arsenate	$Cd_3(AsO_4)_2 \rightleftarrows 3Cd^{++} + 2AsO_4^{---}$	32.7	2.0×10^{-33}
Hydroxide	$Cd(OH)_2 \rightleftarrows Cd^{++} + 2OH^-$	13.93	1.2×10^{-14}
Oxalate	$CdC_2O_4 \rightleftarrows Cd^{++} + C_2O_4^{--}$	7.75	1.8×10^{-8}
Sulfide	$CdS \rightleftarrows Cd^{++} + S^{--}$	28	1×10^{-28}
Calcium			
Arsenate	$Ca_3(AsO_4)_2 \rightleftarrows 3Ca^{++} + 2AsO_4^{---}$	18.2	6.4×10^{-19}
Carbonate	$CaCO_3 \rightleftarrows Ca^{++} + CO_3^{--}$	8.32	4.8×10^{-9}
Fluoride	$CaF_2 \rightleftarrows Ca^{++} + 2F^-$	10.40	4.0×10^{-11}
Hydroxide	$Ca(OH)_2 \rightleftarrows Ca^{++} + 2OH^-$	5.26	5.5×10^{-6}
Iodate	$Ca(IO_3)_2 \rightleftarrows Ca^{++} + 2IO_3^-$	6.15	7.1×10^{-7}
Oxalate	$CaC_2O_4 \rightleftarrows Ca^{++} + C_2O_4^{--}$	8.89	1.3×10^{-9}
Phosphate	$Ca_3(PO_4)_2 \rightleftarrows 3Ca^{++} + 2PO_4^{---}$	29	1×10^{-29}
Sulfate	$CaSO_4 \rightleftarrows Ca^{++} + SO_4^{--}$	5.0	1.0×10^{-5}
Cerium (III)			
Hydroxide	$Ce(OH)_3 \rightleftarrows Ce^{+++} + 3OH^-$	20.2	6.3×10^{-21}
Iodate	$Ce(IO_3)_3 \rightleftarrows Ce^{+++} + 3IO_3^-$	9.50	3.2×10^{-10}
Chromium (III)			
Hydroxide	$Cr(OH)_3 \rightleftarrows Cr^{+++} + 3OH^-$	30.2	6×10^{-31}
Phosphate	$CrPO_4 \rightleftarrows Cr^{+++} + PO_4^{---}$	22.62	2.4×10^{-23}
Cobalt (II)			
Hydroxide	$Co(OH)_2 \rightleftarrows Co^{++} + 2OH^-$	14.89	1.3×10^{-15}
Sulfide	$CoS \rightleftarrows Co^{++} + S^{--}$	24.7	2×10^{-25}
Copper (I)			
Bromide	$CuBr \rightleftarrows Cu^+ + Br^-$	8.28	5.3×10^{-9}
Chloride	$CuCl \rightleftarrows Cu^+ + Cl^-$	6.73	1.9×10^{-7}
Iodide	$CuI \rightleftarrows Cu^+ + I^-$	11.96	1.4×10^{-12}
Sulfide	$Cu_2S \rightleftarrows 2Cu^+ + S^{--}$	47.6	2.5×10^{-48}
Thiocyanate	$CuCNS \rightleftarrows Cu^+ + CNS^-$	12.7	2×10^{-13}
Copper (II)			
Arsenate	$Cu_3(AsO_4)_2 \rightleftarrows 3Cu^{++} + 2AsO_4^{---}$	35.1	8×10^{-36}
Hydroxide	$Cu(OH)_2 \rightleftarrows Cu^{++} + 2OH^-$	19.66	2.2×10^{-20}
Iodate	$Cu(IO_3)_2 \rightleftarrows Cu^{++} + 2IO_3^-$	7.13	7.4×10^{-8}
Oxalate	$CuC_2O_4 \rightleftarrows Cu^{++} + C_2O_4^{--}$	7.54	2.9×10^{-8}
Sulfide	$CuS \rightleftarrows Cu^{++} + S^{--}$	35.2	6×10^{-36}
Iron (II)			
Carbonate	$FeCO_3 \rightleftarrows Fe^{++} + CO_3^{--}$	10.46	3.5×10^{-11}
Hydroxide	$Fe(OH)_2 \rightleftarrows Fe^{++} + 2OH^-$	14.66	2.2×10^{-15}
Sulfide	$FeS \rightleftarrows Fe^{++} + S^{--}$	17.2	6×10^{-18}

TABLE C. 2. (*Continued*)

Compound	Reaction	pK_{sp}	K_{sp}
Iron (III)			
Arsenate	$FeAsO_4 \rightleftarrows Fe^{+++} + AsO_4^{---}$	20.2	6×10^{-21}
Hydroxide	$Fe(OH)_3 \rightleftarrows Fe^{+++} + 3OH^-$	38.6	2.5×10^{-39}
Phosphate	$FePO_4 \rightleftarrows Fe^{+++} + PO_4^{---}$	22	1×10^{-22}
Lead			
Arsenate	$Pb_3(AsO_4)_2 \rightleftarrows 3Pb^{++} + 2AsO_4^{---}$	35.4	4×10^{-36}
Bromide	$PbBr_2 \rightleftarrows Pb^{++} + 2Br^-$	4.4	4×10^{-5}
Carbonate	$PbCO_3 \rightleftarrows Pb^{++} + CO_3^{--}$	13.0	1.0×10^{-13}
Chromate	$PbCrO_4 \rightleftarrows Pb^{++} + CrO_4^{--}$	13.8	1.6×10^{-14}
Chloride	$PbCl_2 \rightleftarrows Pb^{++} + 2Cl^-$	4.79	1.6×10^{-5}
Fluoride	$PbF_2 \rightleftarrows Pb^{++} + 2F^-$	7.4	4×10^{-8}
Hydroxide	$Pb(OH)_2 \rightleftarrows Pb^{++} + 2OH^-$	14.93	1.2×10^{-15}
Iodate	$Pb(IO_3)_2 \rightleftarrows Pb^{++} + 2IO_3^-$	12.55	2.8×10^{-13}
Iodide	$PbI_2 \rightleftarrows Pb^{++} + 2I^-$	8.17	6.7×10^{-9}
Oxalate	$PbC_2O_4 \rightleftarrows Pb^{++} + C_2O_4^{--}$	11.08	8.3×10^{-12}
Phosphate	$Pb_3(PO_4)_2 \rightleftarrows 3Pb^{++} + 2PO_4^{---}$	42	1×10^{-42}
Sulfate	$PbSO_4 \rightleftarrows Pb^{++} + SO_4^{--}$	7.80	1.6×10^{-8}
Sulfide	$PbS \rightleftarrows Pb^{++} + S^{--}$	28.15	7.1×10^{-29}
Magnesium			
Ammonium			
Phosphate	$MgNH_4PO_4 \rightleftarrows Mg^{++} + NH_4^+ +$		
	PO_4^{---}	12.6	2.5×10^{-13}
Arsenate	$Mg_3(AsO_4)_2 \rightleftarrows 3Mg^{++} + 2AsO_4^{---}$	19.7	2×10^{-20}
Carbonate	$MgCO_3 \rightleftarrows Mg^{++} + CO_3^{--}$	5.0	1×10^{-5}
Fluoride	$MgF_2 \rightleftarrows Mg^{++} + 2F^-$	8.2	6.3×10^{-9}
Hydroxide	$Mg(OH)_2 \rightleftarrows Mg^{++} + 2OH^-$	10.95	1.1×10^{-11}
Oxalate	$MgC_2O_4 \rightleftarrows Mg^{++} + C_2O_4^{--}$	4.07	8.6×10^{-5}
Manganese			
Arsenate	$Mn_3(AsO_4)_2 \rightleftarrows 3Mn^{++} + 2AsO_4^{---}$	28.7	2.0×10^{-29}
Carbonate	$MnCO_3 \rightleftarrows Mn^{++} + CO_3^{--}$	10.7	2.0×10^{-11}
Hydroxide	$Mn(OH)_2 \rightleftarrows Mn^{++} + 2OH^-$	12.76	1.7×10^{-13}
Oxalate	$MnC_2O_4 \rightleftarrows Mn^{++} + C_2O_4^{--}$	14.96	1.1×10^{-15}
Sulfide	$MnS \rightleftarrows Mn^{++} + S^{--}$	15.15	7.1×10^{-16}
Mercury (I)			
Bromide	$Hg_2Br_2 \rightleftarrows Hg_2^{++} + 2Br^-$	22.25	5.6×10^{-23}
Carbonate	$Hg_2CO_3 \rightleftarrows Hg_2^{++} + CO_3^{--}$	16.05	8.9×10^{-17}
Chloride	$Hg_2Cl_2 \rightleftarrows Hg_2^{++} + 2Cl^-$	17.88	1.3×10^{-18}
Cyanide	$Hg_2(CN)_2 \rightleftarrows Hg_2^{++} + 2CN^-$	39.3	5.0×10^{-40}
Hydroxide	$Hg_2(OH)_2 \rightleftarrows Hg_2^{++} + 2OH^-$	23	1×10^{-23}
Iodate	$Hg_2(IO_3)_2 \rightleftarrows Hg_2^{++} + 2IO_3^-$	13.71	2.0×10^{-14}
Iodide	$Hg_2I_2 \rightleftarrows Hg_2^{++} + 2I^-$	28.33	4.7×10^{-29}
Sulfate	$Hg_2SO_4 \rightleftarrows Hg_2^{++} + SO_4^{--}$	6.15	7.1×10^{-7}
Thiocyanate	$Hg_2(CNS)_2 \rightleftarrows Hg_2^{++} + 2CNS^-$	19.7	2.0×10^{-20}
Mercury (II)			
Hydroxide	$Hg(OH)_2 \rightleftarrows Hg^{++} + 2OH^-$	25.52	3.0×10^{-26}
Sulfide	$HgS \rightleftarrows Hg^{++} + S^{--}$	52.0	1.0×10^{-52}

TABLE C. 2. (*Continued*)

Compound	Reaction	pK_{sp}	K_{sp}
Nickel			
Carbonate	$NiCO_3 \rightleftarrows Ni^{++} + CO_3^{--}$	8.18	6.6×10^{-9}
Hydroxide	$Ni(OH)_2 \rightleftarrows Ni^{++} + 2OH^-$	15	1×10^{-15}
Sulfide	$NiS \rightleftarrows Ni^{++} + S^{--}$	24.0	1.0×10^{-24}
Silver			
Acetate	$AgC_2H_3O_2 \rightleftarrows Ag^+ + C_2H_3O_2^-$	2.64	2.3×10^{-3}
Arsenate	$Ag_3AsO_4 \rightleftarrows 3Ag^+ + AsO_4^{---}$	22.0	1.0×10^{-22}
Bromate	$AgBrO_3 \rightleftarrows Ag^+ + BrO_3^-$	4.26	5.5×10^{-5}
Bromide	$AgBr \rightleftarrows Ag^+ + Br^-$	12.30	5.0×10^{-13}
Carbonate	$Ag_2CO_3 \rightleftarrows 2Ag^+ + CO_3^{--}$	11.2	6.3×10^{-12}
Chromate	$Ag_2CrO_4 \rightleftarrows 2Ag^+ + CrO_4^{--}$	11.72	1.9×10^{-12}
Chloride	$AgCl \rightleftarrows Ag^+ + Cl^-$	9.75	1.8×10^{-10}
Cyanide	$AgCN \rightleftarrows Ag^+ + CN^-$	13.8	1.6×10^{-14}
Hydroxide	$AgOH \rightleftarrows Ag^+ + OH^-$	7.73	1.9×10^{-8}
Iodate	$AgIO_3 \rightleftarrows Ag^+ + IO_3^-$	7.51	3.1×10^{-8}
Iodide	$AgI \rightleftarrows Ag^+ + I^-$	16.08	8.3×10^{-17}
Nitrite	$AgNO_2 \rightleftarrows Ag^+ + NO_2^-$	3.8	1.6×10^{-4}
Oxalate	$Ag_2C_2O_4 \rightleftarrows 2Ag^+ + C_2O_4^{--}$	10.96	1.1×10^{-11}
Phosphate	$Ag_3PO_4 \rightleftarrows 3Ag^+ + PO_4^{---}$	20.3	2.0×10^{-21}
Sulfate	$Ag_2SO_4 \rightleftarrows 2Ag^+ + SO_4^{--}$	4.80	1.6×10^{-5}
Sulfide	$Ag_2S \rightleftarrows 2Ag^+ + S^{--}$	49.2	6.3×10^{-50}
Thiocyanate	$AgCNS \rightleftarrows Ag^+ + CNS^-$	12.00	1.00×10^{-12}
Strontium			
Arsenate	$Sr_3(PO_4)_2 \rightleftarrows 3Sr^{++} + 2AsO_4^{---}$	18.0	1.0×10^{-18}
Carbonate	$SrCO_3 \rightleftarrows Sr^{++} + CO_3^{--}$	9.96	1.1×10^{-10}
Chromate	$SrCrO_4 \rightleftarrows Sr^{++} + CrO_4^{--}$	4.44	3.6×10^{-5}
Fluoride	$SrF_2 \rightleftarrows Sr^{++} + 2F^-$	8.60	2.5×10^{-9}
Iodate	$Sr(IO_3)_2 \rightleftarrows Sr^{++} + 2IO_3^-$	6.48	3.3×10^{-7}
Oxalate	$SrC_2O_4 \rightleftarrows Sr^{++} + C_2O_4^{--}$	9.25	5.6×10^{-10}
Phosphate	$Sr_3(PO_4)_2 \rightleftarrows 3Sr^{++} + 2PO_4^{---}$	31	1×10^{-31}
Sulfate	$SrSO_4 \rightleftarrows Sr^{++} + SO_4^{--}$	6.49	3.2×10^{-7}
Thallium (I)			
Bromide	$TlBr \rightleftarrows Tl^+ + Br^-$	5.41	3.9×10^{-6}
Chloride	$TlCl \rightleftarrows Tl^+ + Cl^-$	3.72	1.9×10^{-4}
Iodate	$TlIO_3 \rightleftarrows Tl^+ + IO_3^-$	5.51	3.1×10^{-6}
Iodide	$TlI \rightleftarrows Tl^+ + I^-$	7.19	6.5×10^{-8}
Sulfide	$Tl_2S \rightleftarrows 2Tl^+ + S^{--}$	20.3	5×10^{-21}
Tin (II)			
Hydroxide	$Sn(OH)_2 \rightleftarrows Sn^{++} + 2OH^-$	25	1×10^{-25}
Sulfide	$SnS \rightleftarrows Sn^{++} + S^{--}$	25.0	1.0×10^{-25}
Zinc			
Arsenate	$Zn_3(AsO_4)_2 \rightleftarrows 3Zn^{++} + 2AsO_4^{---}$	27.9	1.3×10^{-28}
Carbonate	$ZnCO_3 \rightleftarrows Zn^{++} + CO_3^{--}$	10.68	2.1×10^{-11}
Hydroxide	$Zn(OH)_2 \rightleftarrows Zn^{++} + 2OH^-$	16.7	2.0×10^{-17}
Phosphate	$Zn_3(PO_4)_2 \rightleftarrows 3Zn^{++} + 2PO_4^{---}$	32.0	1.0×10^{-32}
Sulfide	$ZnS \rightleftarrows Zn^{++} + S^{--}$	22.8	1.6×10^{-23}

TABLE C.3. Stability Constants of Some Complex Ions

Compound	Reaction	pK_{stab}	K_{stab}
Aluminum			
EDTA	$Al^{+++} + Y^{-4} \rightleftharpoons AlY^-$	16.13	1.4×10^{16}
Fluoride	$Al^{+++} + 6F^- \rightleftharpoons AlF_6^{---}$	19.84	6.9×10^{19}
Hydroxide	$Al^{+++} + 4OH^- \rightleftharpoons Al(OH)_4^{-}$*	33.3	2×10^{33}
Oxalate	$Al^{+++} + 3Ox^{--} \rightleftharpoons Al(Ox)_3^{---}$	16.3	2×10^{16}
Barium			
EDTA	$Ba^{++} + Y^{-4} \rightleftharpoons BaY^{--}$	7.76	5.8×10^7
Cadmium			
Ammonia	$Cd^{++} + 4NH_3 \rightleftharpoons Cd(NH_3)_4^{++}$	6.92	8.3×10^6
Cyanide	$Cd^{++} + 4CN^- \rightleftharpoons Cd(CN)_4^{--}$	18.85	7.1×10^{18}
ED	$Cd^{++} + 2en \rightleftharpoons Cd(en)_2^{++}$	10.02	1.05×10^{10}
EDTA	$Cd^{++} + Y^{-4} \rightleftharpoons CdY^{--}$	16.46	2.9×10^{16}
Hydroxide	$Cd^{++} + 4OH^- \rightleftharpoons Cd(OH)_4^{--}$	10	10^{10}
Iodide	$Cd^{++} + 6I^- \rightleftharpoons CdI_6^{--}$	6	10^6
Calcium			
EDTA	$Ca^{++} + Y^{-4} \rightleftharpoons CaY^{--}$	10.70	5.0×10^{10}
Cobalt (II)			
Ammonia	$Co^{++} + 6NH_3 \rightleftharpoons Co(NH_3)_6^{++}$	4.75	5.6×10^4
Cyanide	$Co^{++} + 6CN^- \rightleftharpoons Co(CN)_6^{-4}$	19	10^{19}
EDTA	$Co^{++} + Y^{-4} \rightleftharpoons CoY^{--}$	16.31	2.0×10^{16}
Cobalt (III)			
Ammonia	$Co^{+++} + 6NH_3 \rightleftharpoons Co(NH_3)_6^{+++}$	35.2	1.6×10^{35}
EDTA	$Co^{+++} + Y^{-4} \rightleftharpoons CoY^-$	36	10^{36}
Copper (I)			
Ammonia	$Cu^+ + 2NH_3 \rightleftharpoons Cu(NH_3)_2^+$	10.80	6.3×10^{10}
Cyanide	$Cu^+ + 3CN^- \rightleftharpoons Cu(CN)_3^{--}$	28.6	4×10^{28}
Copper (II)			
Ammonia	$Cu^{++} + 4NH_3 \rightleftharpoons Cu(NH_3)_4^{++}$	12.59	3.9×10^{12}
Cyanide	$Cu^{++} + 4CN^- \rightleftharpoons Cu(CN)_4^{--}$	25	10^{25}
ED	$Cu^{++} + 2en \rightleftharpoons Cu(en)_2^{++}$	19.60	4.0×10^{19}
EDTA	$Cu^{++} + Y^{-4} \rightleftharpoons CuY^{--}$	18.80	6.3×10^{18}
Oxalate	$Cu^{++} + 2Ox^{--} \rightleftharpoons Cu(Ox)_2^{--}$	10.3	2×10^{10}
Tartrate	$Cu^{++} + 2Tar^{--} \rightleftharpoons Cu(Tar)_2^{--}$	5.11	1.3×10^5
Iron (II)			
Cyanide	$Fe^{++} + 6CN^- \rightleftharpoons Fe(CN)_6^{-4}$	24	10^{24}
ED	$Fe^{++} + 2en \rightleftharpoons Fe(en)_2^{++}$	7.53	3.4×10^7
EDTA	$Fe^{++} + Y^{-4} \rightleftharpoons FeY^{--}$	14.33	2.1×10^{14}
Iron (III)			
Cyanide	$Fe^{+++} + 6CN^- \rightleftharpoons Fe(CN)_6^{---}$	31	10^{31}
EDTA	$Fe^{+++} + Y^{-4} \rightleftharpoons FeY^-$	25.1	1.3×10^{25}
Oxalate	$Fe^{+++} + 3Ox^{--} \rightleftharpoons Fe(Ox)_3^{---}$	20.2	1.6×10^{20}
Lead (II)			
Acetate	$Pb^{++} + 2CH_3COO^- \rightleftharpoons Pb(CH_3COO)_2$	4.2	1.6×10^4
Cyanide	$Pb^{++} + 4CN^- \rightleftharpoons Pb(CN)_4^{--}$	10	10^{10}
EDTA	$Pb^{++} + Y^{-4} \rightleftharpoons PbY^{--}$	18.04	1.1×10^{18}
Oxalate	$Pb^{++} + 2Ox^{--} \rightleftharpoons Pb(Ox)_2^{--}$	6.54	3.4×10^6

TABLE C. 3. (*Continued*)

Compound	Reaction	pK_{stab}	K_{stab}
Magnesium			
EDTA	$Mg^{++} + Y^{-4} \rightleftharpoons MgY^{--}$	8.69	4.9×10^8
Oxalate	$Mg^{++} + 2Ox^{--} \rightleftharpoons Mg(Ox)_2^{--}$	4.38	2.4×10^4
Manganese			
ED	$Mn^{++} + 2en \rightleftharpoons Mn(en)_2^{++}$	4.79	6.1×10^4
EDTA	$Mn^{++} + Y^{-4} \rightleftharpoons MnY^{--}$	14.04	1.1×10^{14}
Oxalate	$Mn^{++} + 2Ox^{--} \rightleftharpoons Mn(Ox)_2^{--}$	5.25	1.8×10^5
Mercury			
Ammonia	$Hg^{++} + 4NH_3 \rightleftharpoons Hg(NH_3)_4^{++}$	19.4	2.5×10^{19}
Bromide	$Hg^{++} + 4Br^- \rightleftharpoons HgBr_4^{--}$	21.7	5×10^{21}
Chloride	$Hg^{++} + 4Cl^- \rightleftharpoons HgCl_4^{--}$	15.22	1.7×10^{15}
Cyanide	$Hg^{++} + 4CN^- \rightleftharpoons Hg(CN)_4^{--}$	41.5	3.1×10^{41}
EDTA	$Hg^{++} + Y^{-4} \rightleftharpoons HgY^{--}$	21.8	6.3×10^{21}
Iodide	$Hg^{++} + 4I^- \rightleftharpoons HgI_4^{--}$	30.2	1.6×10^{30}
Nickel (II)			
Ammonia	$Ni^{++} + 6NH_3 \rightleftharpoons Ni(NH_3)_6^{++}$	8.49	3.1×10^8
Cyanide	$Ni^{++} + 4CN^- \rightleftharpoons Ni(CN)_4^{--}$	22	10^{22}
ED	$Ni^{++} + 2en \rightleftharpoons Ni(en)_2^{++}$	13.68	4.8×10^{13}
EDTA	$Ni^{++} + Y^{-4} \rightleftharpoons NiY^{--}$	18.62	4.1×10^{18}
Oxalate	$Ni^{++} + 2Ox^{--} \rightleftharpoons Ni(Ox)_2^{--}$	6.51	3.2×10^6
Silver			
Ammonia	$Ag^+ + 2NH_3 \rightleftharpoons Ag(NH_3)_2^+$	7.23	1.7×10^7
Bromide	$Ag^+ + 2Br^- \rightleftharpoons AgBr_2^-$	7.93	8.5×10^7
Chloride	$Ag^+ + 2Cl^- \rightleftharpoons AgCl_2^-$	5.6	4×10^5
Cyanide	$Ag^+ + 2CN^- \rightleftharpoons Ag(CN)_2^-$	21	10^{21}
ED	$Ag^+ + 2en \rightleftharpoons Ag(en)_2^+$	7.70	5.0×10^7
EDTA	$Ag^+ + Y^{-4} \rightleftharpoons AgY^{---}$	7.32	2.1×10^7
Tin (IV)			
Fluoride	$Sn^{++++} + 6F^- \rightleftharpoons SnF_6^{--}$	25	10^{25}
Zinc			
Ammonia	$Zn^{++} + 4NH_3 \rightleftharpoons Zn(NH_3)_4^{++}$	9.06	1.2×10^9
Cyanide	$Zn^{++} + 4CN^- \rightleftharpoons Zn(CN)_4^{--}$	17.92	8.3×10^{17}
ED	$Zn^{++} + 2en \rightleftharpoons Zn(en)_2^{++}$	10.37	2.3×10^{10}
EDTA	$Zn^{++} + Y^{-4} \rightleftharpoons ZnY^{--}$	16.5	3.1×10^{16}
Hydroxide	$Zn^{++} + 4OH^- \rightleftharpoons Zn(OH)_4^{--}$*	15.5	3.1×10^{15}
Oxalate	$Zn^{++} + 2Ox^{--} \rightleftharpoons Zn(Ox)_2^{--}$	7.36	2.3×10^7

* Polynuclear complexes also important.

Abbreviations:

ED = en = ethylenediamine, $_2HNCH_2CH_2NH_2$

EDTA = Y^{-4} = ethylenediaminetetraacetate, $_2(^-OOCCH_2)NCH_2CH_2N(CH_2COO^-)_2$

Ox^{--} = oxalate, $C_2O_4^{--}$

Tar^{--} = tartrate, $^-OOC \cdot CHOH \cdot CHOH \cdot COO^-$.

STANDARD

ELECTRODE POTENTIALS

OF SOME

OXIDATION-REDUCTION

HALF-REACTIONS AT 25°C*

Half-Reaction	E^o, (volt)
$Li^+ + e \rightleftarrows Li$	-3.045
$K^+ + e \rightleftarrows K$	-2.925
$Ba^{++} + 2e \rightleftarrows Ba$	-2.906
$Ca^{++} + 2e \rightleftarrows Ca$	-2.866
$Na^+ + e \rightleftarrows Na$	-2.714
$Mg^{++} + 2e \rightleftarrows Mg$	-2.363
$Al^{+++} + 3e \rightleftarrows Al$	-1.662
$Zn^{++} + 2e \rightleftarrows Zn$	-0.763
$Cr^{+++} + 3e \rightleftarrows Cr$	-0.744
$U^{++++} + e \rightleftarrows U^{+++}$	-0.607
$Fe^{++} + 2e \rightleftarrows Fe$	-0.440
$Cr^{+++} + e \rightleftarrows Cr^{++}$	-0.408
$Cd^{++} + 2e \rightleftarrows Cd$	-0.403

* Some of the material used in this table was selected and adapted from Wendell M. Lattimer, THE OXIDATION STATES OF THE ELEMENTS AND THEIR POTENTIALS IN AQUEOUS SOLUTIONS. 2nd. © 1952. By permission of Prentice-Hall, Inc., Englewood Cliffs, N.J. (The direction of half-reactions and signs used by Latimer are opposite to the one given in these tables.) For extensive tables of standard electrode potentials and their temperature coefficients see Latimer, mentioned above, and A. J. deBethune and N. A. S. Loud, Standard Aqueous Electrode Potentials, Skokie, Ill.: Hampel, 1964.

Half-Reaction	E^0, (volt)
$Tl^+ + e \rightleftarrows Tl$	-0.336
$Co^{++} + 2e \rightleftarrows Co$	-0.277
$V^{+++} + e \rightleftarrows V^{++}$	-0.256
$Ni^{++} + 2e \rightleftarrows Ni$	-0.250
$AgI + e \rightleftarrows Ag + I^-$	-0.152
$Sn^{++} + 2e \rightleftarrows Sn$	-0.136
$Pb^{++} + 2e \rightleftarrows Pb$	-0.126
$2H^+ + 2e \rightleftarrows H_2$	0.0000
$UO_2^{++} + e \rightleftarrows UO_2^+$	$+0.05$
$AgBr + e \rightleftarrows Ag + Br^-$	$+0.071$
$TiO^{++} + 2H^+ + e \rightleftarrows Ti^{+++} + H_2O$	$+0.099$
$Sn^{++++} + 2e \rightleftarrows Sn^{++}$	$+0.15$
$Cu^{++} + e \rightleftarrows Cu^+$	$+0.153$
$AgCl + e \rightleftarrows Ag + Cl^-$	$+0.222$
$HAsO_2 + 3H^+ + 3e \rightleftarrows As + 2H_2O$	$+0.248$
$Hg_2Cl_2 + 2e \rightleftarrows 2Hg + 2Cl^-$	$+0.268$
$Cu^{++} + 2e \rightleftarrows Cu$	$+0.337$
$VO^{++} + 2H^+ + e \rightleftarrows V^{+++} + H_2O$	$+0.359$
$Fe(CN)_6^{---} + e \rightleftarrows Fe(CN)_6^{-4}$	$+0.36$
$Cu^+ + e \rightleftarrows Cu$	$+0.521$
$I_2(s) + 2e \rightleftarrows 2I^-$	$+0.536$
$I_3^- + 2e \rightleftarrows 3I^-$	$+0.536$
$H_3AsO_4 + 2H^+ + 2e \rightleftarrows HAsO_2 + 2H_2O$	$+0.559$
$UO_2^+ + 4H^+ + e \rightleftarrows U^{++++} + 2H_2O$	$+0.62$
$PtCl_6^{--} + 2e \rightleftarrows PtCl_4^{--} + 2Cl^-$	$+0.68$
$O_2(g) + 2H^+ + 2e \rightleftarrows H_2O_2$	$+0.682$
$PtCl_4^{--} + 2e \rightleftarrows Pt + 4Cl^-$	$+0.73$
$Fe^{+++} + e \rightleftarrows Fe^{++}$	$+0.771$
$Hg_2^{++} + 2e \rightleftarrows 2Hg$	$+0.788$
$Ag^+ + e \rightleftarrows Ag$	$+0.799$
$2Hg^{++} + 2e \rightleftarrows Hg_2^{++}$	$+0.920$
$NO_3^- + 4H^+ + 3e \rightleftarrows NO + 2H_2O$	$+0.96$
$V(OH)_4^+ + 2H^+ + e \rightleftarrows VO^{++} + 3H_2O$	$+1.00$
$Br_2 + 2e \rightleftarrows 2Br^-$	$+1.087$
$IO_3^- + 6H^+ + 6e \rightleftarrows \frac{1}{2}I_2 + 3H_2O$	$+1.195$
$O_2 + 4H^+ + 4e \rightleftarrows 2H_2O$	$+1.229$
$Tl^{+++} + 2e \rightleftarrows Tl^+$	$+1.25$
$Cr_2O_7^{--} + 14H^+ + 6e \rightleftarrows 2Cr^{+++} + 7H_2O$	$+1.33$
$Cl_2 + 2e \rightleftarrows 2Cl^-$	$+1.360$
$HIO + H^+ + e \rightleftarrows \frac{1}{2}I_2 + H_2O$	$+1.45$
$PbO_2 + 4H^+ + 2e \rightleftarrows Pb^{++} + 2H_2O$	$+1.455$
$Au^{+++} + 3e \rightleftarrows Au$	$+1.498$
$Mn^{+++} + e \rightleftarrows Mn^{++}$	$+1.51$
$MnO_4^- + 8H^+ + 5e \rightleftarrows Mn^{++} + 4H_2O$	$+1.51$
$Ce^{++++} + e \rightleftarrows Ce^{+++}$	$+1.61$
$Au^+ + e \rightleftarrows Au$	$+1.691$
$H_2O_2 + 2H^+ + 2e \rightleftarrows 2H_2O$	$+1.776$
$Co^{+++} + e \rightleftarrows Co^{++}$	$+1.808$
$F_2 + 2H^+ + 2e \rightleftarrows 2HF$	$+3.06$

ANSWERS TO PROBLEMS

2.1(a). $[Ba^{++}] = 0.20\ M$, $[Li^+] = 0.60\ M$, $[Cl^-] = 1.00\ M$,
$[NO_3{}^-] = 1.00\ M$, $[K^+] = 1.00\ M$.
2.3(a).   HCN, H^+, CN^-; $[H^+] = [CN^-]$; $[HCN] + [CN^-] = 0.20$.
2.4(a). $x = 27$.
2.5(a).   (1) $[H^+] = [X^-] = 0.010\ M$, $[HX] = 1.0\ M$.
(2) $[H^+] = [X^-] = 0.995 \times 10^{-2}\ M$, $[HX] = 0.99\ M$.

3.1(a).   pH $= 14.70$, pOH $= -0.70$, $[OH^-] = 5.0\ M$.
3.2(a). $[H^+] = 13\ M$, pOH $= 15.10$, $[OH^-] = 7.7 \times 10^{-16}\ M$.
3.3(a). $[H^+] = 4.0 \times 10^{-3}\ M$, pH $= 2.40$.
3.4(a). $[H^+] = [ClO^-] = 5.5 \times 10^{-5}\ M$, $[HClO] = 0.10\ M$,
$[OH^-] = 1.8 \times 10^{-10}\ M$, pH $= 4.26$.
3.5(a). $[Na^+] = 0.20\ M$, $[ClO^-] = 0.20\ M$, $[HClO] = 0.10\ M$,
$[H^+] = 1.5 \times 10^{-8}\ M$, $[OH^-] = 6.7 \times 10^{-7}\ M$, pH $= 7.82$.
3.6(a). $[Na^+] = 1.20\ M$, $[Cl^-] = 0.50\ M$, $[CH_3COO^-] = 0.70\ M$,
$[CH_3COOH] = 0.30\ M$, $[H^+] = 7.7 \times 10^{-6}\ M$,
$[OH^-] = 1.3 \times 10^{-9}\ M$, pH $= 5.11$.
3.8(a).   1.00, 1.18, 1.96, 7.00, 10.99, 11.96.

3.12(a). 5.6×10^2.

4.1(a). 3.2×10^{-8}.

4.2(a). 1.00×10^{-5} mole/l, 0.233 mg/100 ml,
$[Ba^{++}] = [SO_4^{--}] = 1.00 \times 10^{-5} M$.

4.3(a). 4.0×10^{-11} mole/l, 2.4×10^{-6} mg/100 ml,
$[Ba^{++}] = 1.2 \times 10^{-10} M$, $[AsO_4^{---}] = 7.9 \times 10^{-11} M$.

4.4(a). $1.0 \times 10^{-8} M$, $1.0 \times 10^{-5} M$.

4.5(a). $[Na^+] = 0.10 M$, $[Br^-] = 0.10 M$, $[Ag^+] = 5.0 \times 10^{-12} M$,
solubility $= 5.0 \times 10^{-12}$ mole/l.

4.6(a). $[K^+] = 0.04 M$, $[NO_3^-] = 0.030 M$, $[I^-] = 0.010 M$,
$[Ag^+] = 8.3 \times 10^{-15} M$.

4.9(a). $5.3 \times 10^{-9} M$, 0.010 mole.

4.10(a). 3.6×10^{-5} mole/l, 6.1×10^{-7} mole/l, 3.2×10^{-9} mole/l.

4.11(a). 5.3×10^{-16} g.

4.13(a). 0.98 to 4.41.

5.1(a). $[Ag^+] = 1.5 \times 10^{-10} M$, $[NO_3^-] = 0.010 M$, $[NH_3] = 1.98 M$,
$[Ag(NH_3)_2^+] = 0.010 M$.

5.2(a). $[NH_3] = 0.18 M$, 0.20 mole/l.

5.3(a). 1.66×10^{-12} mole/l, $[Ag^+] = 1.95 \times 10^{-19} M$,
$[S^{--}] = 1.66 \times 10^{-12} M$, $[NH_3] = 1.00 M$,
$[Ag(NH_3)_2^+] = 3.3 \times 10^{-12} M$.

5.4(a). $1.6 \times 10^{-21} M > [S^{--}] > 1 \times 10^{-23} M$.

5.5(a). $[Zn^{++}] = 0.20 M$, $[Zn(OH)_4^{--}] = 6.2 \times 10^{-18} M$.

5.8(a). 7.2×10^{-5}

6.1(a). 0.56 v, 10^{19}.

6.2(a). 1.6×10^{37}, $[Zn^{++}] = 0.0200 M$, $[Cu^{++}] = 1.2 \times 10^{-39} M$.

6.4(a). 1.74×10^6.

7.1(a). 0.60.

7.2(a). 0.60, 0.37.

7.6(a). $\log K_{sp} = \log \mathbf{K}_{sp} + \dfrac{6.0\sqrt{\mu}}{1 + \sqrt{\mu}}$

9.3(a). $2.2 \times 10^{-3} M$, 0.78.

9.5(a). $6.2 \times 10^{-6} M$.

9.7(a). $5.6 \times 10^{-4} M$, $D = 8$.

9.9(a). $10^{-4} M$.

9.13. 5.7×10^{-3} atm, 3.5×10^{-5} mole/l.

10.1(a). 9.5×10^{-2}.

10.2(a). 9.51×10^{-2}

Index